Essays on

Eusociality & Dissociality

Edited by Joon Yun, MD

ISBN: 978-1-952421-17-4

The editor is grateful for contributions from members of the Yun Family Foundation

Table of Contents

Essays marked with (*) have previously appeared in *Interdependent Capitalism* and *Essays on Inclusive Stakeholding.*

Overview of the Book

Dissociality commodifies our world. As it does, life takes on the feeling of a simulation—not in the sense of Nick Bostrom, Elon Musk, or *The Matrix,* but in the sense that the world feels like a performative version of itself.

Our evolving language helps tell the story. "To act" once meant to do, and now it can mean to perform. "To perform" once meant to act, and now it can mean to pretend. An "image" once meant a picture, and now it can mean a brand.

This type of social evolution is happening across the broader human experience. Campaign promises are performances, social media posts are promotions, telephone rings are solicitations—everywhere we turn, we feel estranged by strangerhood. Alienation abounds.

Behind our masks, we aren't buying what everyone sells. We play along sometimes, but for the most part we have turned off and tuned out. The fallacy hurts, but we are too numb to show it.

It's getting harder to find anyone offering what we would buy: authenticity. That's because the moment for that cultural arbitrage passed as soon as the word "authenticity" became another performative word co-opted by the pretension industry.

Stepping back, we have an intuitive, discomforting feeling that something is off about our lives. Sometimes a specific vignette brings it home, like watching a clown act go after "marks" at a carnival. But mostly, we go about our days with a vague sense that we are living inside something much larger than ourselves.

Yet unlike Jonah of Nineveh, we have no memory of entering the belly of any beast.

How did we get here? When did it all start? Why are we here? What can we do? Where are we going? There's no limit to the number of frames through which we can pursue these questions. That is part of life's beauty.

This book imagines one possible frame for these questions—that the world we experience today represents humanity's dislocation from eusociality to dissociality. We transitioned from living in kin villages, where we were served by those who love us most, to living in a global village, where we are served by those who love themselves most. Long before true authenticity packed its bags, true love had walked out the door.

The journey since has been at times hellish but also divine, odious yet beautiful, backwards and progressive—seemingly all at once. Duality has been our reality.

The reality of this book is that it is two books in one, presented in two parts.

Part One is top down. Comprising the first five chapters, it explores the question of how we got to where we are now through a bioevolutionary frame. From that first-order frame, it examines higher-order social, economic, political, and historical epiphenomena.

Chapter One proposes the first-order root cause behind humanity's shift from eusociality to dissociality: the rise of social entropy. Chapter Two explains dissociality at the individual level: counterparty risk. Chapter Three explores dissociality at the group level: mob mentality. Chapter Four examines what happens when dissocial groups compete: the emergence of self-expanding beasts. Chapter Five probes how dissociality cloaks itself from our awareness: the race to the middle.

Part One approaches these topics from a generalized perspective. As such, should we—our collective humanity—pick up the story ten thousand years from now, Part One would be the more relevant half.

Part Two is more relevant for today. It is more bottom up than top down, more specific than generalized, as illustrations of concepts explored in Part One. It touches on direct experiences of dissociality that we can relate to now.

The three chapters that comprise Part Two are longer than those in Part One. Chapter Six is a collection of experiences we might encounter in everyday life. It offers a representative sample and is by no means an exhaustive account of present conditions. Chapter Seven comprises short stories about the dissocial nature of our stories.

Part Two closes on a more hopeful note. Chapter Eight embarks on a discussion of how to take our dissocial world and invert it back to eusociality. It proposes not a regression back to small kin villages but a path forward toward a eusocial global village. Specifically, it offers ways to leverage the force of present conditions against itself—feeding the beast its own tail, as it were.

Imagine finding our way out of the belly of a beast we were not even aware of being inside of. Cue the scene of Neo and Trinity in *Matrix Revolutions* getting a glimpse of the sun on the last flight of Osiris. Imagine looking around the outside world—decommodified and eusocial—for the first time.

If at any point in the early part of the book you feel discouraged, skip ahead to this final chapter, then return to the earlier chapters later. The spoiler alert in this new rendition of the hero's journey is that we are about to steward the beast's journey homeward. That's when the revolution becomes self-driving. That time is now.

That's not to suggest that there are perfect solutions; if there were an optimum sociality, nature would have already conceived it. Read *this book* not as a destination but as a never-ending series of course corrections toward a receding horizon. The guidepost for making adjustments, always, is to observe whether power serves or subverts—the latter is what we mean by "the beast."

Utopia is not a place arrived at but an idea pursued. In that regard, there will always be work to do and stories to tell. Thank goodness for that. Here's a hat-tip to all those much farther down the road who will clean up our unintended messes and outdo our intended bests.

The book ends with a futuristic Epilogue. While Part One and Part Two deal with the story of human eusociality—the loss of it to date and the potential recovery of it in the future—the essays in the Epilogue expand the notion of eusociality to the universe. Pursuing global human eusociality without including animals, plants, machines, atoms, and space as stakeholders across time adds another layer to what we are already destroying. It's time to tell a new story.

We send you off on that story with a final thought. The unfortunate things depicted herein have largely been wrought not despite but through the good intentions of good people. Just as we intuit that something's behind the discomfort of our present conditions, we also intuit that there's vastly more good will embodied in the human trajectory than it appears. Yet, without eusociality, individual goodness often fails to add up to a great society; that has been our collective failure, and it is our collective opportunity to address it.

It is very much within our power to do better—to make love a true story again.

At the end of the day, the frame presented in this book is merely that. It leaves the rest—the art of consideration—to the reader's infinite imagination.

Definitions

Inclusive Stakeholding (ISH): The philosophical principle of fostering stakeholding in an inclusive manner in social, political, and economic systems as a way to align interests broadly. Inclusive Stakeholding is a theory of change based on the biomimicry of inclusive fitness on larger social scales. Inclusion without stakeholding can lead to balkanization among competing self-interests. Stakeholding without inclusion can lead to the colonization of the excluded. The combined forces of inclusion and stakeholding comprise Inclusive Stakeholding.

Exclusive Stakeholding (ESH): The philosophical principle of non-inclusion of stakeholders affected by the system.

Part One

CHAPTER ONE

First Order: Root Cause

Reader's Guide to Chapter One

We can view history through its social, economic, and political manifestations. Or, we can see these manifestations as symptoms of a common root cause.

Chapter One—the first of the eight book chapters—explores the panoply of historical phenomena through a dislocation in human evolution: the shift of community from eusocial kin villages to a dissocial global village.

The chapter consists of a collection of essays. As is the case throughout the book, some essays are accompanied by an editor's note or extensive footnotes. As is the case here, every chapter is preceded by a reader's guide.

Chapter One's first essay, "Gohyang," *introduces the frame for the book: that we are in the midst of humanity's larger journey homeward. This frame sets even Dorothy's "There's no place like home" mantra in a broader context. Arriving back in her version of reality in Kansas—with no other frame of reference available—Dorothy believes she has returned home.*

But we submit that the reality Dorothy returns to is actually a nested dystopia, masquerading as home. Blinded by our direct experience, it's not easy to sense the totality of humanity's experience—a superorganism that left the cradle of evolution during prehistoric times.

"Kin Skin in the Game" takes us to that imagined cradle—one typically rendered by paleontologists through the bones, artifacts, and genetic footprints that remain.

We instead reimagine life back then through the social realms that we now see at work every day: the altruistic behaviors of family members toward one another. "Social Entropy" explores the incendiary impact of the Promethean Fire on that social structure, which transformed human communities from aligned groups to misaligned individuals. Thereafter, "Exclusive Stakeholding" becomes the operating system of human communities.

"A Brief History of the World" is a recantation of known human history through the lens of this transformation.

Gohyang

There's no place like *gohyang*.

No term evokes a more sentimental response in Korea. The word translates roughly to "hometown" in English. In the Merriam-Webster dictionary, "hometown" further translates to "a place where one was born or grew up." The word may trigger a sense of nostalgia in many Americans, but it is relatively new to the English language, having been coined in 1879.[1] It entered the English vernacular during the 20th century as people began living long enough, and were mobile enough, to move far away from where they were born.

To find a word that better resonates with gohyang, one must trace back to antiquity, to the Latin word *patria*. According to Johan Huizinga, by the time it appeared in the Old Testament, even patria no longer carried the full meaning of the ancient Roman word.[2] It had become an administrative term about location that did not carry a great deal of emotional value.

In the Spanish language, that emotional value is captured by the metaphysical term *querencia*, a spot that a bull stakes out as a strong and safe place during the terror of bullfighting. Ernest Hemingway's *Death in the Afternoon* depicts querencia as "a place the bull naturally wants to go to in the ring, a preferred locality." It is thought of as a home that develops in the brain as one's high ground when all else seems lost.

For our family, gohyang evinces a sense of tenderness beyond a longing for a bygone time and place; it bespeaks of some sacred sanctuary, not unlike a mother's womb, that belongs to all of humanity. To those who have bilingual command of the English and Korean languages, it is evident that there is no commensurate English translation of the word. Like the places where the word originated, the sentiments of hometown and gohyang are worlds apart.

1. See https://books.google.com/ngrams/graph?content=hometown&year_start=1600&year_end=2000.
2. See https://books.google.com/books?id=ge7_AwAAQBAJ&pg=PT105.

These cultures also differ in the way they sense historical timescales. Those who live in the United States are used to thinking of their family history in terms of decades or generations. They are astounded to meet a member of the Daughters of the American Revolution who can trace her ancestry back three or four hundred years.

In societies that have had less migration, however, family histories often can be traced back much further. Members of our family fall into the latter category, having spent time in both our quaint ancestral hometown in Korea and the differently quaint vortex of digital change that is Silicon Valley. We are descendants of General Shin-dal Yun, who was born in 892 AD and served Taejo Wang Geon in founding the Goryeo Dynasty.[3] The exonym "Korea" is derived from the name Goryeo.

As a child in the Yun family, hearing stories about our thousand-year clan history didn't seem real. To a ten-year-old, even a summer feels long, so a thousand years seemed like an unbridgeable eternity. To a fifty-year-old, five decades is a span that flies by in a blink. A thousand years, then, is just twenty of those blinks. Indeed, ten thousand years of recorded human history—the span considered in this book—is only two hundred blinks of an eye in terms of lived human experience.

From this perspective, we can feel a sense of proximity, even intimacy, with our ancestors from a thousand or even ten thousand years ago. Relative to the arc of history, the people we call our ancestors can be seen as our contemporaries who lived in approximately the same brief epoch. We *are* them, and they *are* us.

If the people who lived during the last ten thousand years are our contemporaries, then who were our truly ancient ancestors? What do we know with confidence about their lives, their aspirations, and their fears? For the most part, not much, as we have almost no record of them.

On the other hand, we can surmise a few things. Before humans learned to harness communication, transportation, and energy technologies, they could travel only so far. Thus, we can reasonably assume that prehistoric humans lived in

3. See https://en.wikipedia.org/wiki/Yoon_(Korean_surname); https://en.wikipedia.org/wiki/Goryeo.

social systems composed of extended kin. To our knowledge, they had no means of communicating across distance.

On that single assumption alone, we can build an understanding of what life might have been like back in the prehistoric age.

For you, the reader, our story begins here—even before the time of gohyang.

Kin Skin in the Game

Social species, including humans, exhibit a wide array of social behaviors that affect evolutionary fitness, including those that relate to mating, foraging, survival, defense, communication, cooperation, and competition.

In this book, we focus on behaviors that are displayed when an individual has a vested interest in the success of another individual, such as when they are closely related genetically. Evolution tends to select altruistic behaviors—sometimes even at the expense of one's own welfare—when they are associated with a net-positive outcome from a multilevel selection perspective.[4]

This is called kin selection. The evolutionary fitness model associated with kin selection is known as inclusive fitness. The mathematical model behind kin selection is known as Hamilton's rule: the higher the degree of genetic relatedness, the higher the Darwinian payoff for altruistic behaviors.[5]

The phrase "skin in the game" is often used to refer to situations where we have something at stake in a social or business interaction—in short, we are personally invested in the outcome. Genetics can be a reason for having skin in the game. For example, we are more likely to be personally invested in the outcome of a Little League game one of our own children is playing in than one in which no relations are playing. The same is true of events in which even a distant cousin is competing: we simply are more likely to care than we would in the absence of this relationship.

In this book, we use our own version of "skin in the game" that reflects genetic proximity: "kin skin in the game." Genetically speaking, a parent owns 50 percent of the founders' stock in their children, 25 percent in their grandkids and cousins, and so on.[6]

However, that's not to say that the numbers alone tell the whole story.

4. Wilson, D. S., & Wilson, E. O. (2007). "Rethinking the Theoretical Foundation of Sociobiology." *The Quarterly Review of Biology,* 82, 327-348; https://doi.org/10.1086/522809.

5. See https://en.wikipedia.org/wiki/Kin_selection.

6. The coefficient of relationship is a measure of the degree of consanguinity between two individuals, as defined by geneticist Seawall Wright in 1922.

Mother-to-child, child-to-mother, and sibling-to-sibling relationships may all have the same coefficient of genetic overlap (0.5), but the nature of their nurturing instincts varies considerably. There's clearly more to the story. No doubt, Hamilton's Rule is both a helpful construct yet is incomplete, and kin altruism is increasingly being understood as a complex phenomenon.[7]

Whatever the form, the implications of having bioincentives in the success of others cannot be underestimated. To understand its power in driving evolutionary success, it is worth examining the social behaviors of non-human species that live together in communities exhibiting an exceptionally high degree of kin skin in the game: eusocial species.[8]

Eusociality is the highest level of animal sociality, particularly evident in the Hymenoptera order (ants, bees, and wasps).[9] It is defined by features such as cooperative brood care, overlapping generations living together, and a division into reproductive and non-reproductive groups.

The degrees of cooperation and self-sacrifice among Hymenoptera are remarkable. For example, in *Myrmecocystus mexicanus,* sterile females fill their abdomens with liquid food and hang from the ceiling of the underground nest to store food for the rest of the colony.[10] The queen and king might do all the reproducing while the sterile workers organize into a labor force that supports the overall hive.[11] In some termite species, the soldiers have jaws so enlarged for defense and attack that they are unable to feed themselves and must be fed by workers.[12]

The benefits of such coordinated cooperation are evident in the sheer success of eusocial species. Even though eusociality is thought to have emerged relatively

7. For example, one might wonder if kin altruism could partly operate through a mechanism not dissimilar to what Brian Dias articulated in 2013. In his *Nature Neuroscience* paper, Dias showed that parents who had traumatic experiences paired with specific molecular olfactory sensations passed that olfactory-fear-response association epigenetically onto their offspring. Might olfactory and other forms of consanguinity signalling also be conveyed through lineage in ways that hone altruistic instincts disporporationaly towards one's offspring?

8. Hamilton, W. D. (1996). *Narrow Roads of Gene Land: The Collected Papers of W. D. Hamilton.* Oxford: W. H. Freeman/Spektrum.

9. See https://en.wikipedia.org/wiki/Eusociality.

10. Conway, J. R. (1986). "The Biology of Honey Ants." *The American Biology Teacher,* 48, 335-343.

11. Hölldobler, B. (1990). *The Ants.* Cambridge, MA: Belknap Press.

12. Adams, E. S. (1987). "Territory Size and Population Limits in Mangrove Termites." *Journal of Animal Ecology,* 56, 1069-1081.

recently (around 150 million years ago), eusocial insects account for a disproportionate percentage of the earth's biomass—up to half in some regions.[13] As E. O. Wilson observed, "Social insects are at the ecological center."[14]

The human kin tribe of prehistoric times can also be thought of as a hive of genetic relatives. According to Hamilton's rule, we took care of each other according to our degree of relatedness. We were largely supported by people who had our best interests at heart.

It should also be noted that eusociality as a biological algorithm doesn't scale well as a driver of prosociality beyond the tribal level of human communities. Moreover, not all elements of eusocial societies would be seen as desirable in the modern world. For example, the accounts of vengeful battles between competing eusocial kin tribes reveal extreme levels of dissocial depravity.[15]

That said, updating the biological algorithms of eusociality with new social algorithms of eusociality that scale better remains a profound opportunity for civilization.

Which is where the rest of this story begins.

13. Thorne, B. L., Grimaldi, D. A., & Krishna, K. (2001). "Early Fossil History of the Termites." In T. Abe, D. E. Bignell, & M. Higashi (Eds.), *Termites: Evolution, Sociality, Symbioses, Ecology* (pp. 77-93). New York: Springer; see also http://antbase.org/SISG/sibiodiversity.htm.
14. See http://antbase.org/SISG/sibiodiversity.htm.
15. See https://www.sciencemag.org/news/2018/06/feeding-gods-hundreds-skulls-reveal-massive-scale-human-sacrifice-aztec-capital?fbclid=IwAR1bHmP91U7OHnKgQ1RJIFRhsyhBeZx-4y7UpLrJhJIA9u9nKJ6_Nj9gQvo.

Social Entropy

Family branches become inexorably estranged over the generations.

Put differently, a house divides, and the divided houses keep dividing. The genetic alignment of the original house gets increasingly diluted with each generation, but each new kin hive is internally just as closely aligned genetically as was the original hive.[16] The lineage of Cain and Abel in the biblical narrative was never as genetically close as the brothers themselves. Kin altruism is reborn with every new hive, but as genetic alignment disappears over the generations, competition between hives grows.

The division, dispersal, and genetic fragmentation of kin hives are features of social evolution, not bugs. They promote diversity, intergroup competition, and memetic parallax—the tendency of viewpoints to diverge in groups (to be discussed later). They also help maintain a self-regulated balance between cooperative and competitive instincts in the lineage network. The central house might have stayed put while new branches migrated to new territories, and this balance of in-group attractive forces and outgroup repulsive forces, when viewed at scale, helped maintain the kin-hive structure as mass human migration took hold during the prehistoric age.[17] The human diaspora was under way.

Two things happened subsequently that warrant mention.

First, human dispersal began to reach geographic limits. As humans started to fill various geographic niches around the planet, collisions between tribes with little genetic alignment became more frequent and relative strangers were forced to contend with one another.

Second, humans acquired the so-called Promethean Fire—the knowledge and tools that enabled them to shape their own future.[18] When humans began

16. The resulting structure is fractal in nature, with each genealogical level replicating the proximity of relationships that exist at the next higher order.
17. See https://en.wikipedia.org/wiki/Human_migration.
18. See https://en.wikipedia.org/wiki/Prometheus.

harnessing energy—early on with horses and later through hydrocarbons—it enabled breathtaking progress, but it also increased the probability that competing tribes with little genetic alignment would collide.

The increasing collision of tribes had profound effects, which often were bloody. The diaspora that had started in humanity's cradle began to eat itself, and the slaughter of strangers and hegemony over the vanquished became all too common. On the other hand, the trading of ideas and goods also began to flourish.

Energy increases the entropy of systems, and social entropy is no different. As human social systems harnessed energy, relationship liquidity—defined as the average number of people connected to a particular person and the potential number of transactions among them—vastly increased. There are many beneficial aspects to having a higher number of interpersonal interactions, as having access to more people can increase the probability of finding better partners, in life and in business.

Over time, as converging cultures blended into melting pots, unprecedented levels of prosperity ensued. To be sure, a different force—the centripetal force of migration toward greener pastures of opportunity—also drew us away from each other and into agglomerative forces of human connection, such as cities. As Jane Jacobs chronicled in *The Economy of Cities,* the diversity of people who gathered in the world's first cities came with a diversity of ideas. This dynamic set into motion a positive surge of city inventiveness and city growth that propelled humanity from the Iron Age into the Space Age.

Fast-forward to today, where the social mobility enabled by innovations in technology, communication, and transportation has dramatically increased the liquidity of our relationships. Some of the hardwired social traits we inherited from our tribal-minded ancestors have been rendered maladaptive by evolutionary dislocation and are no longer properly suited to handle modern relationship dynamics.

Our attraction to new social opportunities was shaped when such opportunities were far more limited than they are today. Not unlike our attraction to sweet, fatty, and salty foods, little selection pressure existed in the old world for the evolving upper limit on our attraction to new social opportunities. But in a world where access to new opportunities is virtually limitless, does being intrigued continually by new social opportunities make us happier people?

Whereas our prehistoric tribal predecessors may have had access to a limited set of potential mates, friends, and colleagues, humans in today's mobile society can choose from a vast inventory of possible relationships and even meet distant partners through the Internet. There are many beneficial aspects to increasing the number of transactions. Access to more people, in life and in business, can increase the probability of finding better partners.

On the other hand, increasing relationship liquidity and more possible transactions have resulted in a greater quantity but lower average quality of human relationships. Furthermore, the cost of forsaking existing relationships for new ones has declined. In just the last couple of generations alone, divorce rates and job-change rates have skyrocketed.

Through iterations of social entropy and relationship liquidity, the original pattern of human eusociality—intense in-group cooperation within kin hives and intense outgroup competition between kin hives—has been increasingly displaced by a new type of social structure.

The tribe of one.

Whereas solidarity is a group's unity based on common interests, objectives, standards, and sympathies, solitarity is the state of being alone.[19] The two words are antonyms that help explain the existence of an individual's concurrent and often competing prosocial and selfish urges.

Evolution was once thought to proceed from a solitary species to social ones, but recent phylogenetic studies of bees contradict this assumption.[20] Descendants of an altruistic eusocial group can evolve back to solitary behavior once again—a phenomenon known as a reversal to solitarity.[21] This reversal is thought to have

19. See https://www.merriam-webster.com/dictionary/solidarity; https://en.wikipedia.org/wiki/Solidarity; https://en.wikipedia.org/wiki/Eusociality.
20. See https://www.sciencedirect.com/science/article/pii/S0169534797011981.
21. See https://www.sciencedirect.com/science/article/pii/S0169534797011981.

occurred at least nine separate times during evolution.[22] In some ways, *Homo sapiens* are the tenth known social species to revert from solidarity to solitarity.[23]

Yet the lag error between techno-cultural evolution and our biological factory settings has contributed to millennia of behavior distortions.

22. Gadagkar, R. (1993). "And now...eusocial thrips!" *Current Science,* 64, 215-216; Michener, C. D. (1969). "Comparative Social Behavior of Bees." *Annual Review of Entomology,* 14, 299-342. See also https://www.sciencedirect.com/science/article/pii/S0169534797011981; https://en.wikipedia.org/wiki/Eusociality#Reversal_to_solitarity.
23. See https://www.amazon.com/Social-Entropy-Rising-Relationship-Liquidity-ebook/dp/B07S4F262K/

Exclusive Stakeholding

Exclusive stakeholding (ESH) characterizes the relationship between competing hives. This relationship is, by its very nature, dissocial, and dissociality is the basis of antisocial behaviors. War between eusocial hives is an example.

Today, eusocial hives of our prehistoric past have largely been reduced to nuclear families, and humans are now grappling with how to coexist as a global interconnected hive of self-interested individuals without mutual kin skin in the game. "Game-theory" behaviors replace kin skin in the game as a more dominant modality of communal behaviors.[24]

ESH relies on competition among self-interested individuals and groups to resolve differences. Over the long arc of time, free market competition among ESH groups has enabled enormous prosperity.

But the costs borne along the way by excluded stakeholders have been not insignificant. Over time, ESH in the context of competition, self-interest, and natural inclination tends to produce self-expanding power asymmetries, where the empowered subvert, rather than serve, the disempowered. Examples abound: nepotism, imperialism, nativism, nationalism, colonialism, extractive capitalism, cronyism, corruption, borrowing from the future, pollution, etc. To some extent, the narrative that the world is getting generally better is spoken from the privilege of survivor bias, underwritten by the suffering of excluded stakeholders who paid the price.

Biological evolution and cultural evolution have selected various responses to the self-accumulating nature of power. Senescence is nature's way of setting term limits on power should competition fail to do so. Estate tax, antitrust regulations, and political term limits are examples of cultural technologies that do the same.[25]

24. See https://en.wikipedia.org/wiki/Game_theory

25. Runaway self-accumulation of power deters system evolution. One can consider aging as a prosocial biological trait of robust social systems that enable term limits on power. Estate taxes, antitrust regulation, and term limits are examples of social technologies intended to mitigate self-accumulative tendencies of power. Meanwhile, estate planning and nepotism can be viewed as social technologies intended to counter redistributive forces.

Yet the nature of ESH systems is to continually evolve towards subversion of such cultural technologies.

For example, the system of checks and balances is a preventative cultural technology intended to mitigate self-centralizing tendencies of power. America's founders encoded it into her version of democracy. Yet it is becoming increasingly clear that systems of checks and balances eventually turn into systems of checks written to the account balances of politicians. In other words, systems of checks and balances are undermined by the very force, ESH, that necessitated it in the first place.

All in all, the story of exclusive stakeholding by the empowered, and the price paid by the excluded stakeholders, has been a recurring tragedy of history.

A Brief History of the World

Sometime during the long journey to the advent of the first cities in Anatolia and Mesopotamia roughly ten thousand years ago, the *stewardship* that held tribes together was supplanted by the *leadership* of kings and emperors. Whereas stewardship implies actions taken by a few in the service of the many, the definition of leadership has no such limitations. Indeed, without having kin skin in the game, this new category of human social actor, the "leader," was less incentivized to put the interests of the people before his or her own interests.

All empires and monarchies—from Akkad, Assyria, and Babylon to those of modern times—have been undermined by a lack of incentive structures to replace the inclusive fitness that characterized kin tribes. Corruption, oppression, and self-dealing were the rule, not the exception. Tyranny could be maintained over the generations through the codification of nepotism, which exemplified the perils of inclusive fitness narrowly applied at the expense of collective interests. For quite a while, history was a story of sovereigns ruling over the people instead of on behalf of the people.

Eventually, some societies envisioned governance models in which a chosen few would rule on behalf of the people. Republics emerged in multiple parts of the world, including ancient Europe, the ancient Near East, and Africa.[26] Although a lot of thought went into nurturing these republics, they lacked a critical feature. None of them developed a system of vested interest in the people commensurate to the inclusive fitness of kin tribes.

Abuses not dissimilar to the types observed in totalitarian empires were inevitable. For example, the Roman Republic, which was formed after the Roman Kingdom was overthrown in 509 BC, decayed into a travesty of governance until it finally collapsed in 27 BC with the establishment of the Roman Empire.

By the time Augustus appointed himself the first emperor of Rome—roughly

26. See https://en.wikipedia.org/wiki/Republic.

seven centuries after the city was founded—the ubiquity of self-dealing created an opportunity for cultural arbitrage of epic proportions.

Into that world, a certain future carpenter from Nazareth was born in Bethlehem. By the time that baby was a young man, he had taken to employing parables to remind people to use their power to act on behalf of their dependents. One biblical story recounts how the Nazarene knelt to wash the feet of his disciples in a show of stewardship.[27]

After centuries of nepotistic leaders who gave the world to their sons, a story of a deity who gave his son to the world offered a powerful antithesis.

We know something else: that story caught on.

What made the story stand out was as much what it was *not* as what it was. What it was not was a continuation of thousands of years of accelerating human drift, away from stewardship within kin tribes (where we started) and toward the self-dealing and corruption of monarchs and emperors.

Over time, the magnitude of this story of stewardship began to sink in. However, in the centuries that followed, the transformative promise of that story devolved into a dystopia of disappointment. Just as the Greek and Roman republics had collapsed under the weight of self-dealing, so had the Church gradually done the same as it became the latest institution, lacking sufficient skin in the game in terms of the welfare of its subjects, to rule over instead of on behalf of the people.[28]

To some extent, the original premise of stewardship had been supplanted by the commodified version of itself: performative stewardship. Persecution of Christians by the Roman Emperor had given way to the exploitation of Christians by the Church—until Martin Luther led a revolt in 1517.

One example of self-dealing by the Church was usury. The world of the early Middle Ages was characterized by deflation and zero economic growth. Lending

27. John 13:1-17.
28. Governors of the Roman Empire initially persecuted members of this rising institution. In the year 380 AD, however, Christianity became the State Church of the Roman Empire, halting centuries of official religious persecution.

practices were too often predatory. Perhaps surprisingly, the poor were not the only victims; nobles also forfeited property—even entire estates—to rapacious clerical moneylenders. The Church was powerful, but ultimately it responded to pressure to curb abusive practices within its ranks. By the start of the second millennium of the Common Era, the Church's collapse into self-dealing was so complete that the Pope himself had to intervene. In 1049, Pope Leo IX outlawed interest-bearing loans, declaring at the Council of Reims that "no cleric or layman should be a usurer."

A funny thing happened when Pope Leo IX outlawed usury: trade in interest-bearing loans—now officially outlawed by the Church—took off. With the usurious inclinations of the clergy at least partially restrained, the medieval version of the private financial sector stepped in and opened up a new market space. At the center of this space was a financial innovation known as the bill of exchange, the predecessor to today's bank check. Bills of exchange were important not only because they made it possible for commerce to take place over long distances, which facilitated trade, but also—and importantly—because they constituted a loophole in the papal ban on interest-bearing loans. By carefully designing their terms, the earliest merchant bankers could employ bills of exchange as a way to extend short-term loans. So arose in the history of humanity a new species: the financial services professional.

Within a couple of centuries, the transformative financial service innovation that was the invention of bills of exchange was complemented by an equally transformative legal innovation: the joint-stock company.

The corporation had been born.[29]

The combination of innovations in financial services with those in what we

29. In contemporary corporate law, reference to a joint-stock company is most often considered synonymous with a combination of incorporation (a legal identity that is distinct from that of the shareholders) and limited liability (shareholders are liable for company debts only to the extent of the funds they invested). This is why joint-stock companies are often known as corporations or limited companies.

today refer to as corporate governance had far-reaching and dramatic effects. The very word "corporation"—its root being the Latin *corpus* or body—tells the story. Legal innovation literally created new "bodies" that had their own rights and responsibilities, distinct from human bodies.

However, these paired innovations also created what economists refer to as a principal-agent problem. In this case, the role of the principal is held by the stockholders, while the role of the agent is held by the firm's managers and board of directors (who in theory have a fiduciary duty to shareholders but may be tempted to serve their own interests).[30] In a human body, the brain and the muscles are aligned, and both have an interest in the survival of the individual corpus. In the case of a corporation, the brains (the shareholders) may or may not align with the muscles (the company's management and directors). When this alignment doesn't exist, corporate failure is inevitable.

Thus, not long after its invention, the corporation became the newest institutional vessel for abuse of the people, inflicted by the so-called leaders of the corporation. By the 19th century, the emergence of the factory and large-scale corporate production created a division that defined much of the economics and politics of the 20th century: the division between capital and labor.

The problem was famously framed by Karl Marx, who spoke of the exploitation of a proletarian worker class by the owners of capital, who extracted the lion's share of the value produced and left a bare minimum to the workers. Absent a revolution, the dominant solution to this problem from the workers' standpoint, which emerged in the 19th century, was to organize labor in order to bargain more effectively for an equitable share of the production returns. This new corpus, which sat across the bargaining table from the corporation owners and leadership, was the trade union.

Meanwhile, political parties throughout the industrialized world were adapting to the new realities: one party, usually framed as the conservatives, championed the interests of capital, while an opposing party, usually framed as progressives or

30. The joint-stock company was an innovation that eventually enabled owners to give stock to managers as a way to foster congruent goals.

socialists, championed the interests of labor. From that point on until the present day, relations between labor and capital—which is to say, between trade unions and corporate interests—were (at best) zero-sum battles that often deteriorated into negative-sum tactics, violence, or all-out war.

That is to say, socialists see capitalism as the extraction from the many by the few. Capitalists see socialism as the extraction from the few by the many. What each fails to see is that they both fail for the same reason: the exclusive stakeholding that results in the invisible hand of self-interest entering the cookie jar. Could both systems instead thrive through inclusive stakeholding?

At least one 19th-century economist, William Jevons, had a different idea of how to manage the newly emergent tension between capital and labor. Jevons was one of the three primary architects of the mathematical substructure of modern economics, which is referred to as neoclassical economics to distinguish it from the classical economics of Adam Smith and David Ricardo.

Unlike many of his contemporary economists, Liverpool-born Jevons had witnessed the misery of the factory classes firsthand and was highly sensitive to their plight. "Does not everyone feel that there is an evil at work which needs remedy?" Jevons wrote. "Does not the constant occurrence of strikes, and the rise of vast and powerful organisations of workmen, show that there is some profound unfitness in the present customs of the country to the progress of affairs?" Arguing for the imperative of institutional innovation as eloquently as any modern advocate of equity, Jevons stated, "If the masters insist on retaining their ancient customs; if they will shroud their profits in mystery, and treat their men as if they were another class of beings, whose interests are wholly separate and opposite; I see trouble in the future as in the past."

Jevons continued, "I hope to see the time when workmen will be to a great extent their own capitalists . . . I believe that a movement of workmen towards co-operation in the raising of capital would be anticipated by employers admitting their men to a considerable share of their profits."[31]

31. Auerswald, P. (2017). *The Code Economy: A Forty-Thousand-Year History.* New York: Oxford University Press, pp. 92-93.

A century later, the approach for which Jevons advocated became the new normal. New companies across the United States began to adopt a corporate innovation particularly well-suited to the emerging information age: employee stock options. Issuing stock options promised to accomplish the goal Jevons articulated—the alignment of labor and capital—at the same time it partially resolved the principal-agent problem inherent in the corporate structure. Because a corporation's workers were also its owners, capital and labor would work together. The corporate mind and muscles were aligned.

The effective use of stock options for all employees is one significant element of the mix that made Silicon Valley the entrepreneurial juggernaut it has become and, as a result, has helped to fuel the information age.

At the turn of the millennium, it seemed to many people working in technology that we were on the cusp of a special new era of human history. Alongside institutional innovations like stock options, the invention and growth of the Internet held out the promise of a boundless and inclusive future. The brutal conflict between capital and labor that had defined the ideological battles of the 20th century would give way to a collaborative era in which ownership itself would gradually evaporate in the sunshine of digital equity. Humanity's accumulated knowledge was suddenly available to almost anyone, anywhere. Open-source projects were creating a new paradigm for the world of work—horizontal and distributed rather than hierarchical and centralized. Location would matter less, while talent would matter more.

The endless cycle of institutional whack-a-mole seemed to have come to an end at last. On a free and open Internet driven by mission-aligned teams of creators, the degeneration into self-dealing that had characterized one historical wave after another for two millennia had come to an end at last. Or so we thought.

Little did the Internet idealists foresee a coming era when these companies would be exposed as self-expanding corporate monsters that—as has been the case with every type of institution since the end of the tribal era—grow by self-dealing at the expense of the people.

Silicon Valley companies are now being exposed as the latest institutions, led by so-called leaders, that don't serve the public's interests as well as they serve their own. Clickbaiting to maximize profits and misusing consumer data reflect a company-over-customer philosophy.

Along the way, the highs have been higher, but the lows have also been lower. We've reacted, but none of the solutions we offered quite worked out over time, even if they seemed sound initially. And we've never quite figured out why.

These examples demonstrate a recurring theme: unless first-order alignment issues are addressed, solutions to second-order problems will only create derivative ones. World history as we know it has largely been an epiphenomenon of the central story of our failure to update the bioalgorithms of inclusive fitness in the kin-tribe era with the social algorithms of inclusive stakeholding in the global era.

That is eminently doable.

CHAPTER TWO

Second Order: Dissociality

Reader's Guide to Chapter Two

Chapter Two begins with "Dissociality," a reflection on how the first-order transition from eusociality to dissociality promotes second-order sociopathies. The shift of priorities from other to self is a crucible moment in the human story.

When something is done for the sale value, "Commodification"—a topic explored in the second essay—results. While commodification has raised civilization to new heights, it also casts a long shadow. Perhaps the most devastating commodification is how we use our own powers—the shift from using them to subvert rather than to serve.

Hegemony is not always overt and enforced. "Celebrities" depicts more insidious extractions in which subordinate populations not only buy in to but, through their purchases, also support the power structures. The root cause of "influencer culture"—the displacement of kin by proxy uncles and proxy aunts—is examined further in "Stepmother Effect."

Looking back, the history of dissociality on the one hand has been toward its moderation. As cultural evolution deterred aggression, prosocial rituals such as negotiation and trade ensued. On the other hand, an insidious dynamic looms: the subversion of innate reflexes, such as those regarding taste, mate-seeking, bonding, novelty, status, fear, and even kin altruism itself.

Even cultural tools get harnessed by dissocial power structures to advance their own interests. "Emergence of Language" and "Emergence of Linear-Scale Numbers" explain how language and numbers have served as social technologies but also have subverted human sociality.

Dissociality

The Merriam Webster dictionary defines *dissocial* as "unsocial" or "selfish."[32] *Anti-social* is a synonym. One also can think of dissociality as an antonym of eusociality. Either way, no word better captures the concern of sociality in the modern world.

Selfishness as a concept has many faces. Religious traditions viewed it as a vice. Philosophers such as Bernard Mandeville, Adam Smith, and John Locke saw it as a vice that promotes progress.[33] Ayn Rand claimed it as a virtue.[34] The debate rages on.

But here is a larger question. Can self-interest alone truly protect itself over the long arc against power structures that arise from competing self-interests? In the context of game theory, it is often believed that free will keeps abusive behaviors in check and promotes prosocial behaviors such as reciprocal altruism.[35] In reality, this game is continuously subject to gaming.

In the real world, traits, resources, and good fortune are heterogeneously distributed and competition invariably results in asymmetric power. In eusocial systems, power structures co-serve the interests of individuals and the hive.

In dissocial systems, something different happens: power structures are biomotivated to serve themselves over the interests of individuals, furthering power asymmetries. Ironically, individuals' self-interest inexorably builds exactly the kinds of institutions that Rand feared would oppress individual self-interests.

Furthermore, at the systems level, competition among diverse types of dissocial and eusocial societies favors the selection of perverse forms of the former. One of the lauded features of dissocial structures is the enormous power of innovations unleashed from colliding self-interests. There's no doubt that such innovation can promote good. That said, those innovations often are subverted, thereby accelerating the competition for power among dissocial systems.

32. See https://www.merriam-webster.com/dictionary/dissocial.
33. Mandeville, B. (1970). *The Fable of the Bees.* London: Penguin Books Ltd., pp. 81-83, 410.
34. Nevins, P. L. (2010). *The Politics of Selfishness.* Westport, CT: Praeger, pp. xii-xiii.
35. See https://www.journals.uchicago.edu/doi/abs/10.1086/406755.

Take, for example, the story of America. In the conflict between the eusocial Native American Old World and the technologically advanced dissocial European Old World, the latter brought guns to a knife fight. Those American settlers then declared independence to beat back European imperial powers. That new nation subsequently brought The Bomb to a gunfight, and it became the new imperial power. Power centers within—NYC, DC, Silicon Valley, etc.—began to imperialize domestically, turning middle America into a dustbowl of Walmarts, credit card debt, and addictive behaviors. A nation built on free market competition among self-interests, while making immense positive progress for humanity, had in the process also cannibalized itself.

Vultures joined the feast. External interests, including foreign entities, saw entrepreneurial opportunities within the free market for political favors and disinformation campaigns among profiteering, self-interested commercial institutions.

The exact technologies that had been invented by the people for the people and had done so much good—from the Constitution to nuclear energy to smartphones—were eventually usurped by self-interested parties to serve narrower agendas at the expense of the people.

In other words, every which way you look, self-interest inevitably eats itself.

Commodification

In an idealized world, free market competition drives the creation of products that offer the maximum inherent value at the lowest cost. If the inherent value is rewarded by maximum price earned, the system feeds itself in a self-reinforcing fashion.

The real world behaves a bit differently.[36]

Behavioral economics describes something known as the Winner's Curse, which refers to the tendency for the winning bid in an auction to exceed an item's true worth or intrinsic value.[37] The winner's curse occurs because each participant in the auction has only partial information about the item, so some systematically overestimate its worth and others systematically underestimate it. Naturally, those who overestimate it are more likely to bid high and thus "win" the auction.

Our version of the Winner's Curse comes with a twist. In a world of misaligned incentives, a variety of institutions will compete in the marketplace—think, for example, of banks in the years leading up to the global financial crisis. Some banks might have refrained from adopting financial innovations, such as the packaging and repacking of subprime mortgages as shaky collateralized debt obligations. But other banks jumped in. It didn't take long for the worst practices to "win." Just as "bad money drives out good," according to Gresham's Law, bad institutions drove out good ones in the race to the bottom line.[38] We all know what happened next.

The counterparty risk of bad behavior can be mitigated if social entropy and relationship liquidity remain low. Otherwise, counterparty risk rises. This is where we are today: a world in which people largely produce products, information, and

36. Without kin altruism and reciprocal altruism to protect the mutual benefits of a transaction, extractive transactions can emerge. Such exploitative transactions can occur through coercion, but they can also occur among mutually consenting parties, due to factors such as information asymmetry. In fact, in a corrupted system, the system may even be incentivized to select for greater degrees of information asymmetry over time to maximize the extraction.

37. See https://www.onepetro.org/journal-paper/SPE-2993-PA.

38. See https://en.wikipedia.org/wiki/Gresham%27s_law.

services that prioritize their external value—their trading or sales value—over their inherent value, and without the bioincentives to sustain loyalty.

Commodification is this creation of things for their trading value.[39] To some extent, the desire to produce products and services for trade is a motivator for the kind of innovation and cost reduction we've seen during the history of commerce. In theory, feedback loops such as symmetrical information, rational behavior, etc., enable external value to track inherent value.

This theoretical scenario, however, is often undermined. With low alignment and high relationship liquidity, it's darn near impossible to keep the excesses of capitalism in check. The negative feedback loop that would mitigate violations are not strong enough to deter exploitative behaviors.[40]

Moreover, a culture of commodification stirs a sense of alienation in the public. That difference in feeling when something was done for intrinsic reasons versus extrinsic reasons extrapolates to many dimensions of the human experience.

Even words can be commodified, used for their sale value. One example is the word "leadership." If leadership is such a fundamental need of humanity, one wonders why was the word not coined in English until 1821?[41] In fact, the word "leader" only came into common usage in 1918, and then to describe despots.[42] Did the need for leadership not exist before?

Of course it was needed: the yearning for power to serve the public is a central theme of history.

The harsh reality is that a linguistic masquerade around the word "leadership" enables tyrants to dupe the public and usurp even greater powers. It is thus that, despite more than 50,000 books published with the word "leadership" in the title, the public hungers for the next one.[43]

If leadership is defined as the nurturing of leaders, it is a form of stewardship, and the system then self-propagates the way ecological succession or the cycle of

39. See https://www.marketwatch.com/story/one-crypto-bear-explains-why-bitcoin-is-nothing-more-than-a-trading-sardine-2017-09-19.
40. See https://en.wikipedia.org/wiki/Regulatory_capture.
41. See *Online Etymology Dictionary,* https://www.etymonline.com/search?q=leadership.
42. See *Online Etymology Dictionary,* https://www.etymonline.com/word/leader.
43. See https://www.google.com/search?tbo=p&tbm=bks&q=intitle:leadership&num=10.

life does. However, if leadership is instead defined as the creation of followers—think of Instagram culture—hegemony ensues.

No doubt there are many leaders, even without selfish incentives, who use their powers to serve instead of subvert others. Gresham's Law, however, predicts the rise of the self-serving actors, unchecked by proper incentives, who best delude others, and themselves, that they are doing the right thing for the right reasons.

The commodification of leadership remains a vexing challenge of dissociality.

Celebrities

Picture this. In the historical drama film *Gladiator,* which depicted the Roman Empire as it was heading toward collapse, Maximus, the blood sport's rising star who was challenging the corruption of a philistine society, does a product endorsement.

Between death matches.

For olive oil.

If you think that sounds too absurd, even for a movie about a circus-like era of human history, so did the movie's producers. They deleted this scene from the script because they thought the idea of a gladiator doing commercial endorsements detracted from the realistic feel of the story.[44]

Yet the fact is that gladiators in ancient Rome did use their celebrity to endorse products.[45] The frescos and graffiti of the gladiator era suggest that people back then trusted the purchase advice of their superstar heroes, just as they do today.

There is a powerful scene in the movie when young Lucius approaches Maximus as the latter is about to enter the arena.[46] The poignant moment portrays the influence celebrity athletes wield over children who idolize them. The scene is not dissimilar to that depicted in one of the most successful product endorsements of all time: Coke's campaign in which a wounded Mean Joe Greene is offered a Coke in the tunnel of an arena by a wide-eyed young fanboy, who is thanked with a smile.[47]

In the prehistoric kin tribe era, our larger-than-life role models would have been our aunts and uncles or even our parents. This familial sense is evoked in the Lucius-Maximus scene in *Gladiator,* as Lucius is fatherless and Maximus's own son has been murdered. From an inclusive fitness perspective, idolizing

44. See https://www.ign.com/articles/2000/02/10/not-such-a-wonderful-life-a-look-at-history-in-gladiator.
45. See http://faculty.uml.edu/ethan_spanier/teaching/documents/cyrinogaldiator.pdf.
46. See https://getyarn.io/yarn-clip/20e01580-7afe-4eb5-8fef-5f6e61e65ba0.
47. See https://www.youtube.com/watch?v=xffOCZYX6F8.

and modeling our lives after the celebrities in our prehistoric kin tribe and trusting the ideas and products they endorsed would have served us well. In other words, evolution selected for an innate tendency to trust the endorsements of our avuncular (uncle) and materteral (aunt) heroes.

The avuncular and materteral influences of our tribal past have disappeared from our daily lives and been replaced by transactional strangers who don't have any kin skin in the game. Modern-day celebrities and so-called heroes have an incentive to self-deal and exploit their worshipers for personal gain through extractive capitalism.

Just as soda companies made a lot of money at the expense of children who reflexively and almost unavoidably followed their genetic scripting, trusting the endorsements of modern celebrity influencers is an evolutionary maladaptation.

Stepmother Effect

When it comes to what we receive from others, the change that has mattered most is *who* gives it to us.

Take food, for example. Until the modern age, our food was produced and prepared for us by someone who had our best interests at heart: our mothers. Our moms would never dream of adding something to our food to make us eat more or to save on the cost if there were serious doubts about the effect on our well-being. In the modern world, however, food largely comes from people who care about themselves first and about us second: the food industry. Being a secondary priority is a dangerous position to be in. And yet, here we are.

The food industry, which now serves its own economic interests by using food to bait the public, is incentivized to do things to our food to make us eat more or to save on the cost of producing it, despite doubts about the effects on our well-being. Our food system has shifted from being a high-alignment system during the kin-tribe era to a low-alignment system today.

A million years ago, on the African savannah that is humanity's home, sweets were rare. Our earliest ancestors didn't buy tubs of gummy bears at Costco. They didn't buy orange juice by the half-gallon or consume boxes of breakfast cereal infused with corn syrup. They did, however, need food energy—and lots of it—to satisfy their caloric demands. As a consequence, humans evolved taste receptors, a.k.a. our "sweet tooth," which helped guide us to fruits and other foods rich in nutritionally essential sugars. (We humans are not alone. Think of Winnie the Pooh's love of honey, a fiction based on biological reality: bears crave sugar too.)

Fast-forward to the previous century, when the vast majority of people in industrialized countries and elsewhere sourced foods not through their own hunting and gathering but from a global agro-industrial complex. Processed foods—not just bags of potato chips and cans of soda but also baby formula and energy bars—are engineered to appeal to our ancient cravings for sugar, fat, and salt. Even fresh produce is the outcome of market-driven optimization, with fruit and vegetable varieties bred for appearance, transportability, and, of course, taste. To the delight of our Stone Age taste buds, sugar, fat, and salt sources today are cheap and abundant.[48]

That's just the beginning of the story. The standard process of market competition selects a few winners that work to preserve their market advantage, or industry associations are created to represent large groups of market incumbents—or both.[49] In the case of sugar and corn syrup, this process of industry consolidation has been helped along considerably by a U.S. government policy of protecting industry from market disruptions, which is encoded in a paragon of special interest legislation innocuously known as the Farm Bill. This policy, which stems from the deflationary trauma of the Great Depression, has, over the span of decades, led to the creation of government-sponsored agriculture industry cartels.[50]

What have been the consequences of this political self-dealing and evolutionary impulses run amok? Anyone with even one eye on public health issues in the United States and other industrialized countries in which similar dynamics have played out knows the answer: a deeply disturbing and far-reaching obesity epidemic that is a textbook case of how self-dealing actors can hijack evolutionary inclinations to their own advantage, using institutions (in this case, those of science and representational politics) as their tool.[51] Consumers seem all but resigned to the reality that the food industry will bait them with fake foods—and with fake news about those fake foods.

We have gone from consuming our mom's love to consuming counterparty risk. Seen from a higher perch, this dynamic is nefarious. Once upon a time, our mother or other kin with a vested interest in our success gave us food. Now our food is largely produced by the one who has taken the position that once belonged to our moms but who does not have a vested interest in our success—the

48. See https://www.nytimes.com/2012/06/06/opinion/evolutions-sweet-tooth.html; https://news.nationalgeographic.com/news/2015/01/150118-evolution-flavor-taste-hamburger-ngfood/.
49. See https://promarket.org/sugar-industry-buys-academia-politicians/; https://www.heritage.org/agriculture/report/sugar-shakedown-how-politicians-conspire-the-sugar-lobby-defraud-americas.
50. See https://www.heritage.org/agriculture/report/sugar-shakedown-how-politicians-conspire-the-sugar-lobby-defraud-americas.
51. See https://www.ncbi.nlm.nih.gov/pmc/articles/PMC3228640/.

ever-present self-dealing archetype in *Cinderella, Snow White,* and other folklore tales: the evil stepmother.[52]

In the real world, the vast majority of stepparents do right by their stepchildren. More broadly, humans have adapted far beyond merely following their bioincentives. Everywhere, people routinely mentor, teach, foster-parent, adopt, volunteer, support, and advocate on behalf of others without needing incentives to do so. Kindness among strangers has become part of the social contract.

That said, no kindness compares to the kindness of a mother trying to feed her family. In all cultures, along with the word for "home," the word for "mother" carries the most positive emotional resonance.

Yet the aggregate sum of everyone merely doing good and trying to feed their family, through their participation in the larger labor economy, adds up to a kind of capitalist dystopia few would have imagined.

52. See https://en.wikipedia.org/wiki/Stepmother.

Emergence of Language

Editor's Note: Was the emergence of advanced symbolic systems, including language, both a cause and an effect of dissociality? Low-trust transactions increase the need for negotiation methods, which increases the need for memetic interoperability. Yet even language can be hijacked by power structures to promote their aims.

Evolution is an "as is" phenomenon, and teleology as an explanatory style remains controversial. "Why" questions are generally shunned in science, as they cannot be empirically validated with experimental tools.

With that limitation in mind, trying to explain the evolutionary emergence of human language is even more problematic, given the paucity of direct evidence. Indeed, the Linguistic Society of Paris's prohibition against debating the topic in 1866 remained influential in suppressing formal discourse on the subject in academia until very recently.[53]

Nonetheless, since the time of Darwin, people have speculated as to why language may have emerged during human evolution. Today, varying hypotheses exist as to where, when, how, and why language may have emerged, but there is scant agreement among them.[54] Furthermore, while many have attempted to link the emergence of language with the emergence of modern human behavior, there is little agreement as to the nature of this association. The full list of hypotheses about the putative link are explored elsewhere.

To that list we would like to add another hypothesis. What if the shift from living in kin-based social systems to living in communities of strangers—which lowers genetic alignment and increases counterparty risk—was a forcing function in the acceleration of advanced symbolic systems such as language?

53. Stam, J. H. (1976). *Inquiries into the Origins of Language.* New York: Harper and Row, p. 255.
54. Tallerman, M., & Gibson, K. R. (2012). *The Oxford Handbook of Language Evolution.* Oxford: Oxford University Press.

When genetic alignment is higher, the need to negotiate is lower. One can assume that actions within a community are likely to serve the interests of others and the group. On the other hand, when alignment is low, a greater degree of negotiation is required. Subtleties and nuances matter more, and the precision that comes with naming things, actions, and abstractions enhances the ability of members in a low-aligned community to negotiate more effectively.

Were these burdens a forcing function in the acceleration of human consciousness? Philosophers have long debated the role of language in our ability to experience the world. Gorgias of Leontini posited that the physical world cannot be experienced except through language. Plato and St. Augustine believed that words were merely labels applied to already existing concepts. Immanuel Kant believed language was one of many tools used by humans to experience the world.

In recent times, the theory of linguistic relativity, also known as the Sapir-Whorf hypothesis, has posited that individuals experience the world based on the language they habitually use. For example, because there was a paucity of color terminology in Homeric Greek literature, it is presumed that Greeks at the time did not experience colors as we do today.[55]

Thus, if the shift from high-alignment, kin-based social systems to low-alignment systems was a catalyst for the development of elaborate languages, then the same fundamental shift of our social evolution may also have had a hand in helping us to experience the world in a far richer fashion and to develop advanced thoughts.

In short, the fall from the Eden of our tribal past may have precipitated the development of human awareness, instead of the other way around. (In the standard version of the story, humans were expelled from Eden for eating an apple, which led to awareness.)

A broader interpretation of the theory of linguistic relativity suggests that naming words promotes consciousness through awareness. By transitivity, those who espouse such a view could make the case that the transition of human social

55. See https://en.wikipedia.org/wiki/Linguistic_relativity_and_the_color_naming_debate.

systems from inclusive fitness to reciprocal altruism may have been a forcing function for the emergence of human consciousness.

From the perspective of an individual life, children have no subjective memory of the time before they could name things. Fractally, from the perspective of the species, humans have no recorded memory of the time before the emergence of symbolic systems.

Once such systems emerged, heightened awareness followed as a symptom of dissociality.

Emergence of Linear-Scale Numbers

Many precipitating factors help explain the emergence of advanced symbolic systems during human evolution. We posit that the fundamental shift of human communities—from one predominantly based on kin altruism to one predominantly based on reciprocal altruism—has played a role in the development of advanced symbolic systems.

During the kin-tribe era, we could generally trust that others would act with our best interests at heart without doing an undue amount of due diligence. Today, reciprocal altruism is the dominant form of human transaction, and we now have to count the beans to ascertain whether the transactions are fair. Consuming the counterparty risk of strangers is a far cry from suckling the milk of a mother who loved us unconditionally.

Advanced symbolic systems no doubt increase the evolutionary fitness of social systems based on kin altruism. These systems, however, are even more important for social systems that have to rely predominantly on reciprocal altruism. Counterparty risk is far higher among genetic strangers than genetic relatives. Things and concepts have to be defined and accounted for more precisely when negotiating with counterparties.

A specific example of this hypothesis as it applies to advanced symbolic systems is worth further discussion: the emergence of modern numerical systems based on linear scales. In highly aligned social systems, where numerical information is conveyed on behalf of a relative rather than against a stranger, representation of quantities can be expressed on a compressive scale where there is high resolution at low numbers and low resolution at high numbers.[56] For instance, as a default, a language might only need to contain the symbols for one, two, several, many, and innumerable.

56. See Yun, J. (2015). *Compound Thinking* (Kindle edition); https://www.amazon.com/Compound-Thinking-Joon-Yun-ebook/dp/B015GCMHFC/ref=sr_1_1?ie=UTF8&qid=1547931605&sr=8-1&keywords=compound+thinking+joon+yun.

The benefits of using compressive scales to express quantities are the following. They efficiently represent a wide dynamic range of quantities—the way a Richter Scale is efficient. Second, compressive scales represent quantities according to relevance. When something is scarce, such as bananas in wintertime, being able to discriminate between the quantities one and two matters. On the other hand, there is no fitness relevance to being able to distinguish between a high number of bananas, say 280 and 281. Third, ecological features such as distance to a predator or prey present themselves to our sensory systems on compressive scales. Imagine a line of trees in a forest; trees closer to us appear farther apart from each other than trees that are far from us, and those closer to us will impact our fitness more than those farther away. Finally, everything in nature occurs through a recursion of underlying rules of nature.

Linear scales (equidistance between every consecutive number), however, serve a transactional world better than compressive scales. From the perspective of a counterparty transaction, high resolution at high numbers matters even more than at lower numbers. For example, spending $25,000 for a car versus $24,000 is a huge discrepancy for a customer with a net worth of $30,000; therefore, it is vital that the consumer "senses" the $1,000 difference. A linear scale enables them to sense this far better than a compressive scale.

It is intuitively appealing to speculate that the emergence of linear notational systems for numbers in multiple cultures in the modern world, including India, Assyria, Sumer, Rome, and China, was a function of the rising counterparty risk in transactions associated with the transition from a high-alignment world to a low-alignment world. More specifically, the development of linear scales catalyzed trade among counterparties (reciprocal altruism).

Indeed, the oldest known human counting systems—the small clay tokens invented in the Zagros region of Iran around 4000 BC,[57] the pictographs on tablets representing numerals in 3500 BC,[58] and the abstract numeral system

57. See Schmandt-Besserat, D. (2008, December 8). "Two Precursors of Writing: Plain and Complex Tokens," http://en.finaly.org/index.php/Two_precursors_of_writing:_plain_and_complex_tokens.
58. Schmandt-Besserat, D. (1996). *How Writing Came About.* Austin: University of Texas Press.

used in 3100 BC[59]—are thought to have been developed to facilitate the trade of commodities.[60] Sumerians invented arithmetic soon thereafter, including addition, subtraction, multiplication, and division, to manage their grain trade.[61] As trade flourished, more and more advanced mathematics began to emerge.[62]

There is no doubt that linear mathematics has proven its worth. High resolution at high numbers is the only way to create the kind of precision at high numbers that would allow us to land a rocket on the faraway moon. On the other hand, the nearly complete hegemony of linear-scale mathematics over compressive-scale mathematics also comes with a significant social cost. Natural phenomena occur through the recursion of underlying fundamentals, resulting in compounding outcomes. Linear-scale mathematics does a poor job of capturing and studying compounding phenomena, which happen everywhere in the natural world.

Notice that things in nature curve and that straight lines are man-made. Humans relying on linear-scale mathematics tend to significantly underestimate the compounding effects of both growth and decline, which would be gauged far more effectively on compressive scales.

The cascading downstream effects of the shift from eusociality to dissociality is just one example.

59. Schmandt-Besserat, *How Writing Came About*.
60. See https://en.wikipedia.org/wiki/History_of_ancient_numeral_systems.
61. Nissen, H. J., Damerow, P., & Englund, R. (1993). *Archaic Bookkeeping*. Chicago: University of Chicago Press.
62. See, for example, https://en.wikipedia.org/wiki/Topology.

CHAPTER THREE

Third Order: Dissocial Superorganisms

Reader's Guide to Chapter Three

Groupthink and mobbing are social adaptations that presumably promoted fitness during early evolution. As humans shifted from eusociality and dissociality, however, these traits may have been rendered maladaptive. Chapter Three explores this shift.

"Eusocial Memetics" explores why diversity of thought and creativity flourish when people feel others have their back. By contrast, "Mob Mentality" depicts the opposite. "Escape Panics" and "Middle-Penguin Syndrome" describe the selfish jockeying within a dissocial superorganism. "Metametaphor" shows just how difficult it is for us to overcome herd mentality.

A dissocial power structure benefits when factions splinter and compete against each other rather than against the larger tyranny. "Distrust Bias," like confirmation bias, is an innate brain reflex often subverted by power structures to divide people. Consequences of distrust bias are discussed in "Memetic Parallax" and " Moral Separatism."

"Vertical Civil War" is an articulation of the vexing struggle of individuals against power structures composed of individuals.

Eusocial Memetics

Eusocial *Hymenoptera* colonies with internal genetic alignment exhibit caste polymorphism: drone soldier ants look like twins but look nothing like the queen.[63] This diversity is a feature, not a bug. When there is goal congruence—in organizations, sports teams, etc.—specialization of roles co-serves the interests of individuals and the colony.

Fractally speaking, the human body is also a eusocial colony of genetically aligned cells in which genotypic unity promotes phenotypic diversity. A mucus-secreting goblet cell is wildly different from a photoreceptor retinal cell.[64] By contrast, in a cancer, the loss of alignment of interests among cells promotes competition, dissocial cellular behaviors, and loss of phenotypic diversity—an example of convergent evolution.[65]

Consider memetics through a similar frame. Selection favors memetic superorganisms that sense and process ecological information with high fidelity. Eusociality bioincentivizes high-trust memetic systems that process ecological cues robustly. Memetic monomorphism (groupthink) serves the hive when needed, and polymorphism (viewpoint diversity) also serves the hive when needed.

But human civilization finds itself at an evolutionary crossroads. After eons of genetic arborization that dispersed individuals, the human species has melded into a global, dissocial superorganism. The dissociality bioincentivizes low-loyalty, low-trust memetic systems that motivate sociopathic rather than prosocial use of cues. Trust tilts towards skepticism, and in response to heightened vigilance,

63. See https://www.ias.ac.in/article/fulltext/jgen/076/03/0167-0179.
64. The enormous genotypic *and* phenotypic biodiversity in living systems reflect their common eusocial role of serving as DNA-based replicants in the holobiont. That is to say, at a lower scale of biology, the DNA variation among individual species makes them dissocial and competitive; yet the fact that they all use essentially the same DNA-based backbone for life also makes them eusocial and collaborative at a higher scale of biology. The genotypic variation at the species level can be seen as merely being "phenotypic" physical variations of a common nucleic acid code that links the fate of all living species at the level of a eusocial superorganism.
65. See https://www.ncbi.nlm.nih.gov/pmc/articles/PMC5287060/.

subversive power structures use illegitimate signalling—propaganda, fake news, etc.—to divide the people.

Moreover, power structures are also incentivized to subvert groupthink and memetic polymorphism, co-opting individuals to serve as agents of the beast. Individuals may thereby self-sort into subcolonies of faux foes displaying ingroup cohesion and polarizing outgroup antisocial behaviors.

The current fate of billions of diaspora'ed humans connected to a dissocial hive mirrors the fate of dissocial colonies of prokaryotes living together due to ecological constraints. Selection favored colonies of eusocial biofilms to help them not only survive but thrive. From unicellular biofilms, eusocially capable multicellular superorganisms such as *Homo sapiens* are thought to have evolved.

Can the now-interconnected supercolony of seven billion *Homo sapiens* course-correct toward global eusociality?

Mob Mentality

Editor's Note: Behaviors in dissocial groups differ radically from those in eusocial groups. One example is mob mentality.

In her treatise on the psychology of moral panics, Dr. Maia Newley explores the innate vulnerability of human communities to mass hysteria: even well-educated and apparently rational intellectuals (Hobbes, Bacon, Milton, Locke) were not immune to joining the broad public in becoming convinced that evil witches with great powers were imperiling England.[66]

Such mass hysteria can occur at any scale of social aggregation. For example, Ekbom's syndrome is a delusional disorder in which individuals incorrectly believe they are infested with parasites, and they often compulsively gather "evidence" to present to others.

Pairs of people also can famously exhibit mob mentality. For example, a *folie à deux* is when two people, typically a patient and a spouse, share a delusion and cultivate an "us-versus-them" attitude, which can lead to magnificent triumphs or to Bonnie-and-Clyde-like calamities. Meanwhile, the tendency to fall madly in love can be seen as an adaptive example of a shared delusion, a distributed memetic algorithm selected by evolution—not by Cupid—to temporarily hypnotize pairs to mate before reality sets in, thus propagating the species.

In a *folie à plusieurs,* large groups share a delusion and selectively aggregate the "evidence" that supports their collective belief. Witch hunts fall into this category.[67]

66. See https://www.researchgate.net/publication/276997801_THE_JACOBEAN_WITCH_CRAZE_-_THE_CASE_FOR_FOLIES_A_PLUSIEURS_Psychopathology_of_Early_Modern_Folk_Devils_and_Moral_Panics.
67. See https://www.researchgate.net/publication/276997801_THE_JACOBEAN_WITCH_CRAZE_-_THE_CASE_FOR_FOLIES_A_PLUSIEURS_Psychopathology_of_Early_Modern_Folk_Devils_and_Moral_Panics.

So, are witch hunts a thing of the past? Hardly. Groupthink—including its most extreme form, mob mentality—is an evolutionarily selected feature of living systems that is not going away anytime soon. Moreover, as a consequence of humans' increasing interconnectedness through technology, we are more vulnerable than ever to mob behavior.

First, let's discuss mobbing as a feature. For most of human history, kin skin in the game shared across individuals has enabled humans to coordinate our collective behaviors—brooding, security, resource acquisition, migration, communication, etc.—as a "superorganism" that transcends individual boundaries. In this way, groupthink is an evolutionary win for both the individual and the superorganism.

We would refer to memetic affiliation in prehistoric times as "tribal beliefs"; memetic affiliation today might be referred to as corporate culture or patriotism. Many non-familial groups appropriate the term "family" to foster cohesiveness; it's the instinct that makes the parents of children on sports teams believe that the referees' calls against their own team are unfair. The root is the same: our power to communicate, coordinate, and learn—the essence of our evolutionary advantage as a social species.

As technology increases information liquidity, the role of memetic algorithms in coordinating human group behaviors through all of these categories is rising. On many levels, this is good news. New types of useful communities are made possible by virtue of the technology-enabled connectivity now available.

The problem comes when mobbing tendencies get misappropriated in our dizzyingly global and interconnected world. As eusociality shifted to dissociality, bioincentives of memetic superorganisms became far less aligned with that of individuals. When brains, emotions, and social programs are co-opted, collective consciousness can come under the spell of companies, political groups, foreign nations, social movements, external interests, etc.

In the bigger picture, since eusociality shifted to dissociality, interstrata competition between a superorganism and its individuals has increased, and cooperation between them has decreased. In that state, the superorganism occupies the superposition in the principal-agent relationship. Mob mentality is a manifestation of the vertical principal-agent problem between individuals and the hive.

Can realigning incentives between individuals and the hive deter the misappropriation of mob mentality by memetic superorganisms?

Escape Panics

Editor's Note: This essay discusses the accentuation of groupthink during widespread panics.

Escape panic induces individuals to display maladaptive mass behaviors, such as jamming and life-threatening overcrowding. These physical behaviors have often been attributed to social contagions. Now that information markets and social signaling have moved online, where memetic distribution is both global in scale and instantaneous, a better understanding of escape panics in the digital age is warranted.

Dirk Helbing's article in *Nature* outlined the following features that are characteristic of escape panics: individuals try to move faster than normal; individuals start pushing each other; movements become uncoordinated; clogging is observed at exits; jams cause dangerous pressure to build up; escape is slowed by a build-up of casualties; individuals tend to copy what other people do; and alternative exits are often overlooked or not efficiently used.[68]

A striking example of this last problem is observable in a phenomenon known as the symmetry breaking of escaping ants, an experimentally observable herd instinct triggered by fear.[69] Under normal conditions, a group of ants facing the choice of two equivalent exits out of a box will sort themselves out relatively evenly through both. Under panicked conditions, however, ants facing the same choice of two equivalent exits will crowd toward one, leading to inefficient exit dynamics.[70]

What the experiment reveals is how fear shifts decision-making from direct observation of information to indirect observation through conspecific proxies. Behaviorally, the group appears to shift from a collection of individuals to a swarm. The proximate mechanism has been shown to be a neighbor-imitation behavioral

68. See https://arxiv.org/pdf/cond-mat/0009448.pdf.
69. See https://www.youtube.com/watch?v=6Rz8LNesxGs.
70. See https://www.ncbi.nlm.nih.gov/pmc/articles/PMC4281238/.

instinct triggered by the panic.[71] Aggregation during a fearful incident is thought to be a defense strategy against predation.

As the escaping ants experiment shows, however, this instinctive shift to herd behavior during a fearful incident can be maladaptive. Independent decision-making fails to override the default neighbor-copying instincts, even when the blindly copied behavior is the wrong decision—a decision they would not make if not for their fear.

Parallels in human social behavior abound. The panicked rush toward the exits during a fire can create dangerous stampedes.[72] It can also occur, metaphorically speaking, during high-velocity market crashes, where it manifests as the rush of assets searching for liquidity, which can lead counterproductively to feed-forward evaporation of the latter.[73]

The larger point is this. Blindly copying others' behavior during a panic was a feature of human social evolution during our prehistoric past but, due to evolutionary dislocation, that tendency can be rendered maladaptive. Today, when panics real and imagined can be engineered by self-dealing counterparties (seeking self-promotion and protection from shaming) and disseminated through technology, media, and social media, our inherited tendency to blindly copy the behavior of the herd during a fearful incident can be subverted by outside interests for their own gain.

There is no escaping this reality.

71. To what extent those perceived to have better information—experts—are preferentially followed in nature during escape panics remains to be further elucidated.
72. See https://www.nytimes.com/2003/02/18/us/21-die-in-stampede-of-1500-at-chicago-nightclub.html.
73. See https://www.bloomberg.com/news/articles/2020-03-19/stock-traders-have-no-map-for-the-fastest-ever-bear-market; https://www.cnbc.com/2020/03/23/this-was-the-fastest-30percent-stock-market-decline-ever.html.

Middle-Penguin Syndrome

Editor's Note: Dissociality incentivizes "cover your rear end" behaviors.

Groups of animals fleeing from a predator often exhibit herd behavior. Social behaviors within such panicked herds can vary, depending on many factors.[74] The role that kin skin in the game plays within the herd to determine those herd behaviors cannot be underestimated.

For example, in eusocial herds, individuals may altruistically put their bodies between the predator and the rest of the herd. This can also occur when brood defense is involved.[75]

In dissocial herds, however, what William Hamilton describes as "selfish herd behavior" predominates.[76] In such cases, powerful individuals may compete for the safest inner positions among the herd. Even though the herd may appear to be moving together as a unit, this dynamic emerges from the uncoordinated behavior of self-serving individuals.

From the perspective of any individual, the competitive instinct in selfish herds is to put the bodies of others between themselves and the predators. In the process, the weakest members of the herd are thrown under the Darwinian bus. The selfish behaviors in colonies of Adélie penguins could be understood through this frame.[77] The continuous jockeying for the inner position among the herd can be referred to as middle-penguin syndrome (MPS).

Whereas human social behaviors were shaped during prehistoric times when they lived more eusocially, today's humans live in dissocial communities with low mutual kin skin in the game. As such, one would predict more selfish MPS behaviors among modern-day human herds. This could help explain extractive,

74. See https://en.wikipedia.org/wiki/Selfish_herd_theory#Escape-route_strategies.
75. See http://libres.uncg.edu/ir/uncg/f/O_Rueppell_Review_2012.pdf; https://link.springer.com/article/10.1007/BF00164296.
76. See https://www.sciencedirect.com/science/article/abs/pii/0022519371901895?via%3Dihub.
77. See https://esajournals.onlinelibrary.wiley.com/doi/abs/10.1002/ecy.2823.

cover-your-rear-end behaviors in modern human societies. It would also help explain the near-universal distortion of social, political, and economic institutions that emerge over time, due to self-serving behaviors.

A particular case of MPS has emerged in memetic herds (clusters of memes expressed by individuals) in the age of modern technology and selfish herds.

Selfish social signaling may emerge during periods of widespread panic. Fear, fomented by uncertainty and the cacaphony of information anarchy, can precipitate memetic stampedes. Sentiments can move quickly and pressure can mount for individuals to jockey for safer positions in the center of the memetic herd, lest they be picked off by predatory forces nipping at outlier views. To that end, individuals rapidly copy the views of others.

The predatory forces may be internal, such as shaming by mobs. Whereas mobbing may have been selected by evolution to serve individuals in more eusocial herds, in today's dissocial herds, mobbing may have been rendered maladaptive. Note that founders of classical democracies concerned themselves with mob behaviors that select for demagogues who concentrate power.

Social media is a powerful lever of selfish memetic herding. This new technology has served as a tool for rapid transactions of social signaling, thereby enabling the masses to monitor and adopt herd behaviors. Given the predatory bias toward colony edges, memetic colonies tend to self-organize into protective aggregates. Unlike physical colonies, the center locus of memetic herds can be shared by the masses, which leads to rapid formation of memetic consensus when the level of fear rises. Those with non-conforming views are shamed or excluded from the herd until only a tight, homogeneous consensus survives. Such situations can arise during moral panics and reigns of terror.[78]

The dynamics of the selfish herd continually change and re-form as the threat dissolves. When fear recedes, for example, group behaviors tend to atomize back

78. Selfish herds also exhibit neighbor copying as a form of self-promotion during times of greed. Widespread mimicry in social, political, and economic realms is all too apparent during heady times. Competitive copying of TikTok dance moves as gambits to attain influencer status that can be monetized is just the latest version of a race to the bottom.

into more individual behaviors. The memetic landscape may then return to greater memetic heterogeneity, including more bold challenges to the orthodoxy.

Given the heightened capacity for social signaling in the social media age, the risk of what Friedrich Nietzsche called "herd morality" during times of fear has never been higher. This tendency has become increasingly more apparent throughout the human social landscape. Moreover, while witch hunts have swept through smaller populations at various points of world history—McCarthyism, the Cultural Revolution, and Robespierre's Reign of Terror—the capacity for such hysterias to spread at unprecedented speed and to scale globally has emerged as a significant threat to the future of human sociality.

Metametaphor

Editor's Note: Even without pressure to join the herd, it's harder to avoid than you think.

The story of lemmings jumping off a cliff together is a dramatic, oft-repeated metaphor about the dangers of groupthink—the blind copying of others' behavior. There's just one small problem: it's not true.

The lemmings' Thelma-and-Louise-style legend was born out of a 1958 Walt Disney movie called *White Wilderness* from the studio's True Life Adventures series. It won an Academy Award for best documentary feature.[79] The famous "cliffhanger" scene—which depicted a herd of purportedly on-the-run lemmings heaving themselves into the abyss of the Arctic Sea backed by a score of classical music—was staged.[80]

The truth is that none of the twenty known species of lemmings are indigenous to Alberta, a landlocked province where the film was shot, and none are known to dive off cliffs as a herd.[81] The Alaska Department of Fish and Game tried to repair the impaired reputation of lemmings, but to no avail: the horse had already left the barn, and the lemming myth runs wild to this day.[82]

In a recursive irony, the blind copying of the lemming metaphor to depict groupthink is itself a good example of groupthink. It shows just how much we all live with our heads buried in the sand.

On that point, it is also time to repair the reputation of ostriches. You can find photos of just about anything on the Internet, including people closing the barn door after the horses have left, but you won't find a single photo of an ostrich burying its head in the sand.

79. See https://www.youtube.com/watch?v=xMZlr5Gf9yY&feature=emb_title.
80. See https://www.smithsonianmag.com/smart-news/lemmings-do-not-explode-or-throw-themselves-cliffs-180953475/.
81. See https://www.post-gazette.com/news/science/2003/10/31/We-were-wrong-when-we-blindly-believed-stories-about-lemmings/stories/200310310038.
82. See https://www.adfg.alaska.gov/index.cfm?adfg=wildlifenews.view_article&articles_id=56.

The reason is simple: ostriches don't do that. Once again, it's the human misuse of the metaphor that exemplifies the metaphor.

These are examples of a metametaphor (a portmanteau of *meta* and *metaphor*) that self-referentially embodies the meaning that it depicts. In these cases, they reveal just how strong the tendency is for humans to blindly follow group trends without thinking independently.

Distrust Bias

Distrust bias is the tendency to discount *specific* evidence or ideas—even potentially beneficial ones—if the source is otherwise *generally* considered untrustworthy. Distrust bias is a companion concept to hostile attribution bias, which is the paranoid tendency to interpret benign actions as hostile when the source of the action is considered untrustworthy.[83] As is the case with hostile attribution bias, distrust bias tends to form feed-forward loops: interpretation of actions as untrustworthy increases the sense of distrust. Humans are prone to distrust bias not only at the interpersonal level—we've all seen scorned lovers discredit their ex's every action—but also at the ideological or national level.

Here is an example. On February 1, 1950, the USSR government sent a letter to the U.S. secretary of state sharing the findings of the Khabarovsk War Crime Trials held December 25-31, 1949, in which twelve members of Japan's Kwantung Army were reportedly found guilty of war crimes during World War II.[84] In 1950, the USSR also published in English a large volume of official documents related to the trial, including verbatim testimony and documentary evidence.[85]

However, in the context of the Cold War and growing distrust of the USSR, the United States dismissed the trial and its findings as communist propaganda until the 1980s, near the end of the Cold War.[86] The West's dismissal of the Khabarovsk trial illustrates the influence ideology has on our ability to make fair cross-ideology assessments.[87] The book the USSR published that contained verbatim testimony and documentary evidence has long been out of print, but it resurfaced on Google Books in 2015.[88]

83. See https://en.wikipedia.org/wiki/Hostile_attribution_bias
84. See https://www.archives.gov/files/iwg/japanese-war-crimes/select-documents.pdf at page 51; https://en.wikipedia.org/wiki/Khabarovsk_War_Crime_Trials.
85. See https://books.google.com/books?id=ARojAAAAMAAJ&printsec=frontcover#v=onepage&q&f=false.
86. See https://link.springer.com/article/10.1007/BF02448905.
87. See https://link.springer.com/article/10.1007/BF02448905.
88. See https://play.google.com/books/reader?id=ARojAAAAMAAJ&hl=en&pg=GBS.PA10.

Today, uncivil ideological cold wars are on the rise everywhere, due in part to technologies that amplify human psychological foibles that favor and pay attention to polarizations. Those who attempt to point out the merits of an idea that has been developed or espoused by an ideological opponent are branded as sympathizers. Such shaming is a behavioral feature of mobs, and it empowers the memetic superorganism to silence dissident memes and assert control over individual free expression. To blind ingroups to outgroup memes, a memetic superorganism may brand these non-conforming memes that leak through memetic boundaries as conspiracy theories, thereby shoring up memetic boundaries.

To some extent, memetic immune systems mimic biology. In our bodies, a system of antibodies tags entities perceived as untrustworthy, and a second system of macrophages deletes the tagged entities from the system. In our minds, one layer of the memetic immune system can tag information that threatens the power structure as untrustworthy, and another layer of the memetic immune system can delete those tagged memes from the system. Seen through a dystopian lens, one might conclude that this is how memetic superorganisms will keep human brains washed of suspected impurities.

The consequences of the distrust bias driving ideological polarization in the age of technology could be enormous and potentially disastrous. Specifically, the tendency is for an ideological group to adopt a position opposite to that of their ideological enemy, regardless of the position's merit, under the assumption that whatever position the latter adopts must be a bad idea and thus must be opposed. Anyone falling short of opposing every position the enemy takes could be named a sympathizer and be cast out from the ingroup by the ideological mob.

Power structures may harness emerging technologies for memetic control. In the human body, the electrictrification-based nervous system prosocially coordinates a trillion eusocially interconnected cells. By contrast, in the body politic that is the digital age, electrification of the Internet dissocially coordinates billions of interconnected humans, amplifying their pre-digital-age sociopathy.

The resulting oppositionalism is self-escalating. The attack by memetic superorganism *A* toward memetic superorganism *B* triggers a fear response in the target memetic superorganism, which increases its power to control internal individualism through mobbing. This allows memetic superorganism *B* to sharpen its weapons and counterattack against memetic superorganism *A,* which triggers fear in superorganism *A* and gives it greater control over its own group through mobbing. This cycle is mutually reinforcing and promotes polarization.

Heightened mutual suspicion can spill over into other domains, not unlike a nuclear chain reaction: the radioactive energy of one topic is the input that triggers the nucleus of another topic to split in two, setting off a cascade. What results is a mutation that turns human sociality into something more cancer-like.

Who would want this to happen?

For clickbait media companies or commercially motivated political fundraising machines that profit from polarization, the resulting social pollution is merely an externality—not unlike environmental pollution. It's not something they always want to see happen, but willful blindness is a powerful force.

Foreign governments that *do* want to see this happen—whose aim is to undermine another nation—play the game at a higher level. They can foment the memetic fragmentation and internal combustion of another nation merely by paying to subvert the same profiteering instinct of clickbait media companies and commercially motivated political fundraising machines—or any mercenary social, political, or economic institution.

In summary, all that is needed to take down civilization is natural inclinations, free markets, and exclusive stakeholding—one company not having a properly aligned stake in the nation's success, one nation not having a properly aligned stake in another nation's success, and so on.

Memetic Parallax

What comes immediately after "world peace"? Civil war.

Humans tend to form groups that disagree with each other, just as a single cell must divide through the process of mitosis to enable the growth of an organism. Polarization isn't a bug of human social software. It's a feature.

From a multilevel selection perspective, evolution selects for systems with features that increase evolutionary capacity: imperfect replication, sexual reproduction, predation, programmed death, memetics, etc.[89] Add to this list memetic parallax, which is the tendency of meme groups to diverge into competing views.[90]

The tendency toward groupthink driven by memetic algorithms can trap systems in local minima in the adaptive landscape that impede evolutionary novelty and constrain fitness. From a multilevel selection perspective, a certain degree of trait diversity (meme or physical) and conflict within a system promote that system's robustness and evolutionary capacity. The propensity to disagree with others—many incentives and disincentives influence the process—can promote overall system fitness.

Indeed, at a macro level, the theory of evolutionary capacity predicts that selection favors systems that generate feature parallax (diversity over consensus), especially when it promotes conflict to a degree that optimizes competition and accelerates evolution.

For example, predator-prey competition can be viewed as a feature parallax that increases evolutionary capacity for both parties. In cultural evolution, the fragmentation of sociopolitical-economic philosophies such as socialism and capitalism represent ideological parallax that benefits the overall system.

Sexual reproduction is another system that promotes genotype and phenotype parallax (male vs. female). The competing agendas promote vast increases

89. Wilson, D. S. (2015). *Does Altruism Exist? Culture, Genes, and the Welfare of Others*. New Haven, CT: Yale University Press.
90. Socratic dialogues and Hegelian dialectic both reflect the fundamental role of memetic parallax in human inquiry. See Hegel, G. W. F. (2018). *The Phenomenology of Spirit*. Oxford: Oxford University Press.

in novel features, including cognition and social behaviors. Sexual reproduction is an example of active, positive selection for traits among gender counterparties with highly aligned interests in a joint venture (i.e., offspring) in which each owns a 50 percent vested interest (or less for males, given paternal uncertainty).[91]

When there is trust, respect, and alignment (genetic and otherwise) between individuals or groups, the parallax can be highly beneficial to the individuals and to the combined superorganism. A second opinion from a trusted partner is a significant value-add in partnerships, teams, and marriages. Without trust and respect, memetic parallax can lead to dysfunctional polarization, including war.

A fundamental feature of diversity is often overlooked. When there is common ground (inclusive stakeholding), diversity can be mutually beneficial. On the other hand, diversity without common ground (exclusive stakeholding) can lead to separatism.

When aggressive Balkanization occurs, individuals and groups also tend to unify internally, to merge, form meme alliances, and cooperate—again, many incentives and disincentives influence the process—to compete against outside groups. When facing a common external enemy, we tend to settle our internal disputes and unite against the enemy until a larger consensus (peace) results.[92]

Once the enemy is dissipated, memetic parallax reemerges. The victors of WWII entered peace discussions with a spirit of international cooperation and exited with divergent agendas. There's wisdom in the notion that a war defending humanity against a common enemy such as an alien invasion would diffuse existing ideological stalemates, at least until the victory parade.

When a group Balkanizes into many groups, an important second trait of social systems deters further fragmentation—the hive mentality. From a total-system evolutionary-capacity perspective, there is a tradeoff: parallax is preferred

91. The standard explanation for why sexual reproduction evolved is increased diversity. We propose a more salient reason: that vested interest in our offspring is a powerful driver of the careful and motivated selection of our reproductive partner's traits.

92. The rise of nationalism following the unification of warring factions in Japan instigated imperialism—a pattern repeated often around the world when clans settle their differences in favor of common aims. Imperialism and colonialism are outlets that feed the beast and keep internal harmony within warring tribes.

over consensus, but continued parallax and over-fragmentation are also costly. "Us-versus-them" campaigns require retaining and growing the "us." Social systems are wired to consolidate and grow their membership and their power as superorganisms. That said, under certain conditions, meme groups can atomize considerably to the point of information anarchy.

Selection favors dynamic systems capable of both consensus memetics and memetic parallax in a context-dependent fashion. A fractal is apparent. Evolution favors systems with capacity for two features that are themselves in competitive parallax: the tendency for groupthink and the tendency for parallax.

Thus, the arcs of human and evolutionary history are poised to teeter between competition and cooperation in a never-ending vacillation across the different scales of biology and time. Peace will follow war, and war will follow peace. And so on.

In other words, parallax can also occur along the temporal axis, rather than only at any one moment in time. The Hegelian dialectic—and to some extent the entire intellectual history of Marxism—is built on this. Dualities that evolve over time—the cyclical procession of thesis, antithesis, and synthesis—delineate the progress of human society. Now, of course, the absurdity of Hegel was his announcement that the early 19th century was the termination of this historical procession of thesis, antithesis, and synthesis.

Others have tried to make similar claims. Francis Fukuyama wrote in *The End of History and the Last Man* that the neoclassical liberal state was the end of history.[93] All of that is juxtaposed against the capitalist model, where progress is the persistent story of heroic individuals overcoming adversaries. But neither Marxism nor capitalism can lay total claim to the progress of humanity.

But it doesn't really matter if this is the end of history, if this is the final synthesis, or if this is just the next stage. The point is that inclusive stakeholding is the imminent next thing that, as an intertemporal memetic parallax, stands in stark contrast to the exclusive stakeholding of history.

93. Fukuyama, F. (1992). *The End of History and the Last Man.* New York: The Free Press.

Moral Separatism

Editor's Note: Exclusive Stakeholding of the moral high ground is a basis of humanity's atrocities. It enables dehumanization and demonization of other stakeholders.94

We have a labradoodle puppy. He is descended from a lineage of canines that hunted and killed for their meals. He harbors a dog-eat-dog past—in a way, we all do—but he lives in a dog-eat-dog-food present. He's been bred and domesticated to be friendly, furry, pettable, and even hypoallergenic; he definitely has not been bred for his ability to hunt. He couldn't hurt most creatures if he tried. If he started eating what he killed, we'd probably have to give him away. His benign nature is the reason we didn't keep the receipt when we brought him home.

That's not to say that the killing has stopped. He is eating dog food made from living creatures that were killed out of sight by "the machine." "The machine" is probably one of those self-expanding beasts that profits from a system we feed with our purchases.

What about humans? Are we dog-eat-dog people, or are we good, innocent people who bask in the illusion of domesticated bliss while displacing the "evil" onto a system that we disparage but nonetheless feed into (and thus are a part of)?

By and large, we have chosen the latter narrative. Rather than accepting our total identity—that each of us is simultaneously the beast and the hero, the good and the evil, the problem and the solution—we have chosen, through memetic parallax, to segregate these identities. We have chosen to believe in the myth of separatism.

In separatism, we identify ourselves as the good people. We allow ourselves to be dissociated from the harm we cause others. Meanwhile, we identify the opposing side as bad people. To that bad team we assign the role of monsters, beasts,

94. On the other hand, ISH of the moral high ground is a path to more peaceful, less destructive solutions.

legendary creatures, and the devil, all the while being blind to the possibility that we too might have a beastly side. We are blind to the fact that the other side sees us as the monsters. We fail to see that the "machine" is merely the sum of other people trying to feed *their* families.[95] We are both good and evil but find it difficult to accept that duality.

Instead, we war.

95. Systems tend to dissociate people from the externalities caused by their actions. Meat eaters don't mentally link their consumption to an out-of-sight horror chamber that is the industrial slaughterhouse, and online shoppers don't associate their clicks with creating pollution through the modern-day consumer-industrial complex. That the cumulative effects of everyone's instinct to feed their family—the inherited kin skin in the game "good" instinct—feeds the "bad" beast of dissociality in interconnected communities is one of the ways beasts subvert individual good intentions.

Vertical Civil War

The earlier section, "Mob Mentality," conceptualized the existence of an invisible super-superorganism that promotes competition among moral superorganisms, which are themselves composed of memetically programmed dissocial humans.

In effect, this represents competition between the strata of biology: groups against individuals, supergroups against groups, super-supergroups against supergroups, etc.

Traditionally, evolutionary competition is contemplated horizontally: individual against individual—group against group—but competition can also be contemplated vertically across strata. Put another way, what's good for the goose may not mean it's good for the geese—the interests of an individual in a group are pitted against that of the group, and so on.

In eusocial hives, however, there is also a high degree of alignment of interests, not only among individuals in the hive but between those individuals and the hive superorganism. Strata are sometimes incentivized to collaborate.

It is thus that there are always incentives to compete and to collaborate (coopetition) between different strata of biology. In other words, selection and evolution also operate along the vertical dimension in the hierarchical organization of biological strata.

Biology has multiple strata, including molecular, organelle, cellular, histological, organismal, tribal, species, and ecosystem levels. Each stratum of a hierarchy can be viewed as comprising a collection of one or more "living" units, and these units can compete and collaborate horizontally within each stratum.

Collaboration and competition also occur in the vertical dimension *between* the different strata. That the fates between strata are collaboratively aligned is self-evident: the fate of a cell in the body is generally tied to that of the whole body and vice versa. However, the fates of units in different strata are often not aligned due to interstrata competition.

Here is an example. Apoptosis of a particular cell in the body (self-death) typically helps the strata above (the organism) and the strata below (genes, from an inclusive-fitness perspective). Even as its adjacent strata benefit, however, the individual cell is dead.

Here is another example. Diversity in a system tends to reduce its fragility and increase its robustness. The same diversity, however, also tends to increase stress between underlying members of the system. Since the system itself, as well as individual members that comprise the system, are all simultaneously trying to achieve self-stability against stress, the system is vertically competing against the interest of the strata below it (the strata of individual members of the system).

During the 20th century, this type of framing was depopularized by those who opposed the notion of "group selection." To some, the framing of selection operating at the level of the "selfish individual" was supreme. This view was consistent with the broader rise of the culture of individualism in all fields of contemporary thought, from psychology to political philosophy. Later, Richard Dawkins offered a competing view of the "selfish gene." Farther downstream, David Sloan Wilson and E. O. Wilson offered a more integrated model through their multilevel selection theory.

One way to apply that framing here is to appreciate that the single mutation of sociality—the loss of mutual kin skin in the game among individuals in a group—shifted human hives from eusociality to dissociality. Thereafter, the degree of collaboration between the hive superorganism and the individuals in the hive decreased, while the degree of competition between them increased.

In the former eusocial scenario, the system tendency is for the power of the hive to be continually redistributed to the individuals in the group. In the latter dissocial scenario, the system tendency is for the power of individuals to be continually hijacked by the superorganism to maintain its tyranny over those very people. Calls or movements to decentralize power—independence declarations and blockchain—are merely symptomatic band-aids that fail to resolve (instead of revolve) the root cause: the shift from eusociality to dissociality.

Tyranny by the superorganism (the system) is not executed top-down per se; it occurs bottom-up through free markets and natural inclinations. This is what makes it so insidious: humans are subverted, not served, by the aggregate sum of their own self-interests operating on autopilot. This is the history of the world, writ large.

What's happening here is a civil war that no one can see. The civil wars we are familiar with are those fought horizontally: one set of colored coats pitted against another.

But the most insidious civil war is the one that pits individuals against institutions—which are themselves made up of individuals. In the end, it's a tragic war of people against themselves.

Think of the most important civil war as the one being fought—now raging for thousands of years of dissociality in the post-eusocial era of human evolution—in the vertical dimension: individuals versus superorganisms composed of individuals. It's a fight we didn't even know we were fighting, and any raging war that you're not aware of is one you are losing.

What would it mean for people to win this vertical civil war? To attack the very people we are fighting on behalf of is a horror; to attack the beast is to detonate a suicide bomb on ourselves.

Think of it in a different way. To win this vertical civil war is to liberate people from the superorganisms that hold them emotionally, memetically, intellectually, mentally, socially, and, most importantly, institutionally self-captive.

The key word here is self-captivity. No chains are binding and no passports have been revoked. It turns out that our duty never was to attack the beast or even to rescue people on the other side. We were the "other" all along. There never was a beast, nor was there ever a hero. We had domesticated ourselves into serving the power structures through our aggregate self-interests and free markets.

We were the very people we were trying to rescue.

CHAPTER FOUR

Fourth Order: Stampede of Self-Expanding Beasts

Reader's Guide to Chapter Four

Selfish cancer cells extract resources from neighboring eusocial cells. Competition among cancer cells exacerbates the dynamic. This phenomenon is explored at the societal level in Chapter Four.

"Big Man, Rubbish Man" offers an introductory anthropological frame—the contrast between trade-based societies that grow or die versus stewardship-based eusocial societies that are more stable.

The more generalized form of this frame is explored in "Warlordism's Org Chart," which helps to explain why dissocial groups composed of transactional mercenaries continually mine for value elsewhere to feed the inner self-expanding beast.

"Race to the Bottom Line" explores the larger game played among self-expanding beasts. When the entire system is composed of barely held-together clusters of mercenaries, the competition for loot among mercenary teams creates a "Stampede of Self-Expanding Beasts"—the title for an eponymous essay.

Competitive imperialism is the inevitable outcome of this stampede. While the traditional forms of imperialism are better known, the lesser known forms flying under the radar are worth mentioning here.

"Domestic Imperialism" depicts the beast's tendency to self-cannibalize. Moreover, the tip of the spear finds the most vulnerable, valuable, domestic targets. That's the subject matter behind "Mrs. Doubtfire, Only More Seductive."

There are even more invisible forms of imperialism. "Intertemporal Hegemony" brings to light the intertemporal imperialism and colonialism of excluded stakeholders in the time dimension rather than in the spatial dimension. "Moral Imperialism" makes the case that cancerous behaviors have no theoretical or theological limits in an infinite universe.

Where is all this going?

Robust systems exhibit self-stability through negative feedback loops whereby perturbations to the system tend to self-consume themselves. One might expect dissociality to similarly self-correct. To some extent it does, through disruptive revolutions that address the symptoms, if not the causes, of power asymmetries.

But empirical observations tell a different story. Not only do those revolutions invariably fail to cure the first-order cause, they also do not occur frequently enough relative to the average human lifespan to matter for most. In the meantime, instead of self-correction, what we observe are nefarious feed-forward loops similar to cancer, and competition among them; over time, the most aggressive cancer cells within the tumor dominate the tumor.

If we zoom out to a much larger frame, we can see the yin and yang cycle at work. "Dissociality Supercycle" is a fractal of the Big Man cycle playing out at a civilization level—told through the story of the Bronze Age collapse.

Ultimately, even all the positive intentions of prosociality become undermined by the very thing it was trying to overcome: self-interest. Without addressing the root cause—not updating the bioalgorithms of eusociality with social algorithms for the interconnected age—good ideas and innovations become co-opted by dissociality.

It is thus that history, one way another, becomes a historic recursion, in which a narrow group of stakeholders—certain families or circles of power—compete with each other to extract from the rest of the stakeholders until the process comes to an end, an epiphenomenon described in "Historic Recursion."

As an outro, the last essay of the chapter reimagines how to think about the so-called peaks and valleys of civilizational histories: "Are Dark Ages and Enlightenments Reversed?"

Big Man, Rubbish Man

Editor's Notes: A hallmark of dissociality is conditional loyalty. When conditions are no longer met, power structures decay. The "Big Man" social system of Melanesia can be seen as a microcosm of the rise and fall of societies built on conditional loyalty.

Moka is the reciprocal gifting of pigs, a ritual that determines social status among people in Mount Hagen, Papua New Guinea. The ritual inspired the anthropological notion of the gift economy, which was studied by anthropologist Marshall Sahlins as a way to distinguish between the transaction-based "Big Man" social system of Melanesia and thc kin-based "Chiefly" social systems of the Polynesian islands.[96] The comparison is not dissimilar to Robert Trivers's notion of reciprocal altruism versus William Hamilton's notion of kin altruism.

The Moka system is a dyadic exchange relationship wherein there is a delay between gift and counter-gift—thus establishing a temporary debt. Without the debt, there is no relationship. If the debtor returns a gift of equal amount, then the exchange is simply reciprocity and the exchange ends.

Alternatively, the repayer can add Moka to the repayment gift to increase their prestige—that is, to establish a reputation as a Big Man—and put the receiver in debt. Whereas a debt fully paid ends further interaction, ongoing renewal of the debt relationship sustains the relationship.

Meanwhile, a failure to repay in full pushes a person's reputation toward the other end of the scale, the "Rubbish Man."

People prefer to give gifts to Big Men, who are likely to repay extra Moka, than to Rubbish Men, who are unlikely to repay fully. Over time, this competition

96. There was no distinction between social, political, and economic institutions before the era of legal institutions, as they acted in a holistic fashion within a community. An economic interaction had social or political effects and vice versa. See https://en.wikipedia.org/wiki/Karl_Polanyi.

between high-status men, each of whom tries to give bigger gifts to the other, inevitably results in a hierarchy of power.

The Big Man system is based on the ability to persuade, rather than on the ability to coerce. As such, it is inherently unstable and eventually collapses.[97] This is because the ritual is ultimately constrained by the land's capacity for the pigs and the recipient's ability to delegate the pig husbandry.

When the Big Man is no longer able to source extra Moka, he fails to repay a gift with Moka and his status declines. This creates a chain reaction in which the network of men no longer have extra Moka to pay their underlings, thus destroying the whole power structure. Then everything starts all over again.

The only way to defer this downward spiral is to pursue a grow-or-die strategy, which necessitates imperial gambits to an outer world that will generate enough Moka to feed the inner power structure.

By contrast, the Chiefly system of resource redistribution among Polynesia's island communities—Hawai'i, Tonga, and Fiji—is based on kinship.[98] Lower status kinsmen are obligated to be loyal and pay tributes to higher status kinsmen, who then re-allocate those resources to the group. The Chiefly system is considered much more stable than the Big Man system.[99]

These contrasting systems—Chiefly and Big Man—can be thought of as archetypal systems that represent eusociality and dissociality, respectively. Big Man systems are more competitive internally and more imperialistic externally—retaining the conditional loyalty of mercenaries requires meeting their rising expectations—than Chiefly systems of eusociality, where loyalty is kin-based.

The tendency to compete internally and pursue imperialism allows Big Man systems to accrue more innovations than Chiefly systems. Thus, when the two systems collide, the Big Man systems, with their more advanced innovations and weapons, tend to prevail. That is why Big Man systems have survived cultural selection.

97. Sahlins, M. (1963). "Poor Man, Rich Man, Big Man, Chief: Political Types in Melanesia and Polynesia." *Comparative Studies in Society and History*, 3, 294-297.
98. Sahlins, "Poor Man."
99. Sahlins, "Poor Man."

While they are in power, Big Man civilizations tend to create substantial cultural output. However, these dissocial civilizations are inherently unstable, risking collapse back into decentralized eusocial tribes.

The rise and fall of all power structures can be contemplated through this frame.

Warlordism's Org Chart

Editor's Note: In dissocial groups, loyalty is bought, not inherent. That is the foundation of mercenarism. In a system of competition among mercenary groups, a race to the bottom line ensues.

A stewardship-based org chart can be depicted as a downward-pointing pyramid, with the more experienced staff positioned at the bottom tip. But you probably haven't seen one like that. The more familiar org chart has the pyramid tip pointing upward: a president stands atop the team's shoulders, while those on the bottom compete to climb the corporate ladder. This seemingly minor difference reveals a lot more about the world than just the nature of org charts.

For most of human evolution, loyalty to one's group was based on mutual kin skin in the game. After the great diasporas, however, humans began living in collective societies of genetic strangers. Without mutual kin skin in the game to sustain cohesion, loyalty became a tradable commodity among these groups.

What emerged is a system of warlordism—mercenaries playing a game of power (along the lines of game theory) among semi-stable transactional organizations. In this kind of regime, it is natural for members to conceive the org chart as a pyramid to climb. Think about why people who become president are congratulated by their friends on social media rather than being consoled for the increase in responsibility: in a system of mercenaries, the presidency is too often sought for the spoils of power.

To compete with each other for members, mercenary organizations develop a system of rewards, punishments, and myths—part of a toolkit of institutions to retain and grow membership. The competition for mercenaries among groups fuels the continuous race-to-the-bottom nature of these tools.

Compensation is one example. Warlordism necessitates a system of mutually escalating yearly raises to retain the fickle services of restless mercenaries—agents of fortune whose loyalties are to the royalties of the almighty bounty, and not to the almighty him- or herself. In all systems of warlordism—from the empires, knights, pirates, and samurais of bygone eras to the multinational conglomerates and nation-states of today—the ever-rising bidding for services of mercenaries creates a "grow-or-die" dynamic.[100]

Therein lies the seed of competitive imperialism. The drive to seek bounties elsewhere is driven not only by romantic adventurism but by the practical requirements of keeping one's team together. If the warlord doesn't give a raise, then a competing warlord management team down the block will. A competitive race to the top among warlords sparks a competitive race to the bottom for resources to feed the inner restless beast of self-escalating raises.

The competitive imperialism among warlords managing bands of mercenaries is the dynamic behind what Dr. Eric Weinstein of Thiel Capital calls "embedded growth obligations." The same market forces that impelled Julius Caesar to expand his empire also impel modern-day imperialists to innovate new lines of business: they need to keep harvesting resources from elsewhere to keep their own line workers marching forward and marching together. Meanwhile, the world pays the price (while reaping some benefits) for this continuous hunt for bounty among competing warlords.

That is to say, the same ruthless forces of competition that promote useful innovations also corner bands of mercenaries into a double bind: in a world of warlordism, one can either competitively harvest resources from the larger system to feed the insatiable beast within, or one can be a victim of someone else's warlordism. This is how the cancer of imperial warlordism grew from the cradle of modern civilization, metastasized unsustainably all over the planet using heavenly bodies to navigate their locations, and now is planning to navigate toward those heavenly bodies in rockets. Even the low inflation target of the Federal Reserve—in a world where innovation and depopulation are injecting deflationary

100. See https://www.amazon.com/Grow-Die-Unconditional-Biologic-Trade-off-ebook/dp/B07GNSF4LD.

forces—can be understood as a contrivance needed to promote the "grow-or-die" ethos in the system of mercenaries.

The implications of this warlordism dynamic have affected every dimension of the human experience. Starting as early as age two, future line workers in the industrial complex today funnel themselves into an "assembly-linesque" indoctrination system, shed any hint of a happy childhood, and begin the hypercompetitive rug-rat race for admittance into prestigious schools and entry-level jobs at prestigious mercenary houses. Most won't come even close to this kind of outcome and will toil in other types of trenches.

Warlordism also creates the up-and-out dynamic of labor as talent advances in age. In a system of rising yearly salaries, when the utility of an employee hits a peak inflection point and begins to fade with age, a regression of salaries doesn't fit the narrative. Thus, individual careers fractally mirror the "grow-or-die" tendency of the organization. Senior executives are thus promoted out of the firm and put out to pasture.

It is thus that the standard org chart with the pyramid pointing up is a pyramid scheme of power that fuels the competitive imperialism across all social, economic, and political institutions—driving humanity bonkers.

Race to the Bottom Line

Why doesn't the bad behavior of the food industry get selected out? Can't we call out bad behaviors, regulate them, shame them, and make them go away? Shouldn't self-interest balance out other competing self-interests?

Maybe.

But what we see everywhere instead is a race to the bottom of commodification. In a market of producers vying to sell, they maximize for trading value. Whereas we might hope that the maximization of trading value is associated with maximum inherent value, delivering inherent value is generally at least somewhat more costly. Competition, if anything, selects for lowest costs, which—in expected value terms at least—will be associated with lower inherent value. We therefore have a setup where the system selects for maximum commodification at the lowest cost, which is associated with the lowest inherent value.

Today, the behavior of commodification is evident in every type of institution; it's a race to the bottom line.

But it's not just institutions against the people. It's also people against the institutions. Students face increasing pressure to learn not for the inherent value of learning but for the external value (commodification) that effort carries in competing for college admission.[101]

It's also people against each other. People too often engage in activities not for the inherent purpose but for the promotional purpose of signaling to others. Virtue was once pursued for its own sake; by contrast, virtue signaling is the commodification of virtue.[102] Cars were once bought for transportation; now, the signaling value of automobiles often exceeds the functional value. Clothes used to clothe a person; now fashion is about signaling. People used to live life for itself; they now too often attend events or travel for the purpose of promoting their

101. Some might make the argument that the commodification of learning is a good thing, in that a certain portion of the population would not have studied as hard as they did had that effort not been made for the sake of gaining college admission.
102. See https://en.wikipedia.org/wiki/Virtue_signalling.

personal brand on social media. None of us needs to work hard to come up with examples; in the 21st-century performative version of real life, these phenomena are anything but subtle.

Commodification has replaced the authentic original purpose everywhere where there are relationships between strangers. Accounting used to serve the purpose of creating insights into one's family-owned business. Now accounting is often misused to tell fictional truths to sell stock to a stranger.[103] Language might once have been used predominantly to convey truth, but now it is far more often used to sell a story.

In the end, everyone in this carnival is someone else's mark.

And in the end, bad behavior drives out good behavior until only bad behavior is left. The system self-selects for extractive and exploitative institutions. If we force one media company to use less clickbait, another will use more in order to pick up the other's market share. If we force one food company to use less sugar, another will use more to fill the void in the market. In a way, the Kardashians and high-fructose corn syrup are really the same phenomenon—the inevitable outcome of a race to the bottom line, when misalignment meets unbridled capitalism.

No matter how many times we rerun this simulation called society, the system will eventually select for simulators: purveyors of fake news about fake heroes endorsing fake foods.[104]

The above examples of race-to-the-bottom-line incentive structures are hardly exhaustive. Even the media comes out looking like a paradise of aligned incentives when compared with the perversity of the U.S. healthcare system. It suffices to Google "opioid epidemic" for an indication of the litany of misaligned incentives that plagues the industry.[105] Marx claimed that religion was "the opiate of the masses"; today, tragically, opiates are the religion of the masses.

103. Barach L., & Gu, F. (2016). *The End of Accounting and the Path Forward for Investors and Managers.* Hoboken, NJ: John Wiley & Sons.
104. See https://www.forbes.com/sites/brucelee/2016/10/03/spotting-fake-celebrity-endorsements-of-diet-and-health-remedies/#1a8f7696fdb2.
105. See https://www.nytimes.com/2018/10/11/nyregion/doctors-charged-opioid-prescriptions.html; https://www.amazon.com/Dreamland-True-Americas-Opiate-Epidemic/dp/1511336404.

Indeed, virtually every institutional category is to some extent similarly undermined by perverse incentives. The purpose of this book is not to catalogue them all. We are merely using the cited examples across disparate industries to focus on the higher-level meta-observations.

Looking across the entire landscape of institutions, the race to the bottom line is neither a purely corporate illness nor an affliction of the most selfish and greediest among us. Rather, it is a structural condition affecting every institutional category in society: human social institutions have consistently exhibited a tendency to evolve from high-minded, high-alignment origins to self-dealing, low-alignment maturity. As Harvard evolutionary biologist Joseph Henrich notes in *The Secret of Our Success,* all pro-social institutions collapse over time under the pressure of self-interests.[106] [107]

Each generation and each category of institutions we have described—across the broad sweep of history in Chapter One and across the landscape of society in Chapter Two—did not start out with the intention of self-dealing and racing to the bottom line. Yet, in every instance, from the Roman gladiators endorsing olive oil to the pharmaceutical industry marketing opiates, self-dealing and corruption have been the outcome. Why is that? Were the pioneers who led the institutional revolutions we have described inadequately virtuous? Did they attack the beasts of inequality and injustice with inadequate zeal?

Obviously not. In social systems as much as in biological ones, an individual virtue or failing is no match for evolutionary pressures.

Ever since humanity set out on the long march of progress that has lifted us from hunter-gatherer subsistence to today's urbanized abundance and material

106. See https://press.princeton.edu/titles/10543.html.
107. One of the perils of prosociality is the intertemporal separation of near-term benefits from the long-term harms of good intentions. It is thus kindergarten feels innocent but high school feels a bit like a John Hughes movie; sugar tastes good now but the medical bills later make you feel sick; the Internet feels utopian at its birth but dystopian 25 years later. Indeed, whereas eusocial systems are biomotivated to adopt technologies and prosocial intentions in a way that benefits the group and individuals, dissocial systems are biomotivated to harness technologies and prosocial intentions in a way that benefits the power structures. By the time people realize that the latter is happening, they often find themselves being in too deep to pivot easily. Yet the self-consuming nature of prosociality doesn't stop people from getting on the next prosocial merry-go-round.

bliss, we have been seeking to create or re-create new versions of incentive systems that we left behind: the natural kin-skin-in-the-game alignment that came from our life in kin tribes.

Put differently, for tens of thousands of years, we have been trying to get home to the place where our moms, dads, cousins, uncles, and aunts cooked our meals and were our heroes. That is the world our reptilian brains expect when we wake up every morning. That is the world our reptilian brains believe we walk into every day. And time after time—whether we have awoken in Babylon or Brooklyn or whether we have set out in golden raiments or Gore-Tex parkas—we are disappointed to find that the institutions we trusted did not have our best interests in mind.

Stampede of Self-Expanding Beasts

The race to the bottom line naturally produces a world in which the institutional clusters that gain market dominance are pressured toward ever-increasing scale and reach. Having exhausted all the naturally occurring inputs to growth, the systems naturally evolve to generate false needs and artificial demand.

The product of the healthcare system is not healthcare but old people who need more healthcare. Big Food produces fake news about fake foods that increases the demand for more fake food. The product of the media is ignorance and confusion, which creates a demand for more media.

Any institution that has self-engineered runaway feedback loops in its favor (and at the expense of the people) will hereby be referred to as a "self-expanding beast." In a race-to-the-bottom world, the system selects for the beasts that are the best at self-expansion—that is, the worst actors—by crowding out less malignant beasts.

This is essentially how a colony of cancer cells operates.

Here's how a self-expanding beast feeds its own growth. Although our medical industry has made amazing breakthroughs that extend life, at the end of the day—and at the end of our days—doing so without addressing aging has produced an older, sicker population that needs more healthcare. An aging population needs and pays for more healthcare, which expands the balance sheet of healthcare companies. These companies in turn use the balance sheet for more development and marketing.

This vicious cycle has become a self-expanding beast. Based on current trends, according to the *Annals of Family Medicine,* the average cost of healthcare for an American will surpass the average American's income in 14 years.[108] One would hope for a healthcare system that keeps people so healthy that it puts itself out of business. Instead, what we have is a system that will put people out of business.

108. See https://www.ncbi.nlm.nih.gov/pmc/articles/PMC3315130/.

The sum total of all well-meaning physicians taking care of patients somehow adds up to something different than an ideal system.

Similarly, the more consumers feed on clickbait news and drink high-fructose, corn-syrup- flavored drinks, the more they empower the very companies that are exploiting them. The self-dealing media and food companies in turn get richer and are able to spend more dollars to bait consumers, creating a vicious cycle. In essence, these companies turn into self-expanding beasts that grow ever larger at the expense of the people. Growth is the dominant imperative of these beasts: the greater their capacity to self-feed, the greater their capacity to grow.

Yet, if anything, we seem to be falling in love with our captors as if afflicted with Stockholm syndrome.

Our modern love affair with institutions could be traced back to the publication in 1651 of *Leviathan,* Thomas Hobbes's masterwork of political philosophy. In *Leviathan,* Hobbes famously describes the dog-eat-dog reality of human existence in the absence of political communities of various types:

> In such condition, there is no place for industry; because the fruit thereof is uncertain: and consequently no culture of the earth; no navigation, nor use of the commodities that may be imported by sea; no commodious building; no instruments of moving, and removing, such things as require much force; no knowledge of the face of the earth; no account of time; no arts; no letters; no society; and which is worst of all, continual fear, and danger of violent death; and the life of man, solitary, poor, nasty, brutish, and short.[109]

Given that, when Hobbes wrote these words, the homicide rate in Europe was about ten times what it is today and life expectancy was about half, it may seem a bit surprising that Hobbes would draw any favorable contrast at all between his era

109. Malcolm, N. (2012). *Leviathan*. Oxford: Oxford University Press.

and preceding ones. But Hobbes was one of the first major figures of the English Enlightenment, and he wrote at the very beginning of a remarkable era of social and economic transformation. Along multiple irrefutable dimensions, our lives today are dramatically better—not only than those of our distant relations who lived in kin tribes but also than the contemporaries of Thomas Hobbes.

So what's the problem?

The French existentialists had it partly right when they pointed to the "alienation" of modern man. Compared to our ancestors, we modern humans are systematically disconnected—alienated, if you will—from our neighbors, the natural world around us, our leaders, the food we eat . . . pretty much anything but our pets and the success of our children, both of which have become the subject of almost unbounded attention and investment.

However, alienation is a passive term that does not capture the active misalignments of interests that are the focus of Chapter Two. Institutions in a race-to-the-bottom-line world evolve into self-expanding beasts in a purposeful, incentive-driven process.

It is evident that diaspora/social entropy/trying to coexist with each other in low-alignment communities is inherently fraught with issues. Our factory setting is biased toward exploiting strangers (reciprocal altruism) and trusting kin (kin altruism). Furthermore, reciprocal altruism has a flaw; it assumes that both sides have insight. In reality, information asymmetry, as well as hardwired preferences (attraction to porn, violence, sugar, gossip), can be mined. Moreover, the feed-forward nature of the growth of these self-expanding beasts allows them to invest in better ways to increase and exploit asymmetries.

If you take a step back and look around, you see a stampede of self-expanding beasts rising everywhere on the horizon. The healthcare industry creates its own demand. Big Food, Big Media, and Big Tech self-expand through cycles of addiction. Big Banks, Big Tobacco, and Big Government are also growing at the expense of the people.

Meanwhile, humanity has been brought to its knees.

Domestic Imperialism

Editor's Note: Imperialism isn't merely directed toward external stakeholders in distant lands. Internal colonization of domestic stakeholders also occurs.

News of international trade has been back on the front page. Few, however, think about trade at the local level as a fractal microcosm of international trade. It is critical to understand the flow of goods and services in exchange for money across the imaginary membranes of America's small towns. For many of them, participation in the global economy has been decidedly one way: consumption without production. The social, economic, and political implications cannot be underestimated.

For illustrative purposes, imagine there was a small town that produced widgets. It excelled at producing widgets and was able to "export" them across the town membrane to other towns and, eventually, into the global marketplace. The town "imported" a basket of goods from other towns and from the global markets. The total value of the widgets exported equaled the total value of the basket of goods imported. The tax dollars sent across the membrane to local, state, and federal treasuries equaled the tax spending coming into the town in the form of public schools, public hospitals, infrastructure projects, government jobs, etc. Finally, the town had local mom-and-pop businesses that circulated all those dollars around the town. The town's total balance sheet remained in a steady state.

Then, one day, the widget manufacturing was outsourced to China, where labor was cheaper, and Walmart came to town, bankrupting all the local businesses. After a while, the whole town shopped at, and worked for, Walmart.

A trade deficit formed across the imaginary membrane that surrounded the town. Of the dollars Walmart collected from the townspeople, it paid a portion back to them in the form of salaries. The rest of the dollars crossed the town membrane and went out into the world. Some of the dollars went to Walmart's suppliers in China, some were sent as tax dollars to Washington, DC, and some were sent as profits to Walmart headquarters in Arkansas. With every passing year, as products came into and dollars left the town, the town's balance sheet dwindled further.

After a few more years, the town's balance sheet went into the negative, and the townspeople began using credit cards to borrow money from New York bankers. There was no reason for young people to wait around the town, as they only stood to inherit debt, so they moved to New York to work for the banks. Those remaining turned to the opioids that were brought into town to soothe the pain, and even more dollars went out across the membrane.

Later, Amazon put Walmart out of business, and all remaining working townspeople were laid off.

Whether it's across municipal boundaries or social classes, domestic imperialism is the phenomenon of self-expanding economic institutions that extract from within their own nation. Now, go back and read those front-page stories about international trade again.

Mrs. Doubtfire, Only More Seductive

Editor's Note: In a system of exclusive stakeholding, economic power structures colonize and cannibalize the weakest excluded groups in the system. This story is a cautionary tale.

With the Ring acquisition and the launch of Amazon Key, Amazon now offers video-monitored package delivery inside the home.[110] Walmart, in partnership with August Home, is offering a similar service, whereby delivery personnel monitored by cameras will place groceries in your refrigerator.[111] Viewed through a wider lens, however, there may be more to this story than meets the eye.

What appears at first glance to be a monitoring service for consumers may also be a monitoring *of* consumers, sparking privacy concerns. The video footage—plus what the delivery personnel can see—puts these companies in a *prime* position to gather valuable data inside your home or refrigerator, on top of what they gather from smart consumer devices in your home.

But delivery services may also be trying to get a foot in your door for an entirely different reason. The in-home delivery service may evolve into an interactive in-home sales service. While this may sound implausible today, let's think this through.

From a retailer's perspective, the point-of-sale continues to shift from physical stores, where valuable high-touch human sales interfaces are possible, to the largely impersonal world of e-commerce. As the digital sales process becomes increasingly commoditized and competitive, retailers are motivated to find new ways to differentiate and strengthen their communication with consumers.

Already, some personnel who previously used to work strictly as service providers are starting to hawk products at the point of service. For example, roadside service providers now sell new batteries in addition to jumping cars.

110. See http://money.cnn.com/2017/10/25/technology/business/amazon-key-delivery/index.html.
111. See http://money.cnn.com/2017/09/22/technology/future/walmart-home-delivery-groceries/index.html.

Those kinds of service relationships, by their nature, are one-offs. However, what about recurring relationships that happen inside the home, where trust is the most valuable commodity? In-home service providers of all types—maids, plumbers, electricians, nannies, tutors, caretakers, nurses, gardeners, cooks, and delivery personnel—are suddenly looking to retailers like untapped sales channels. Imagine an Amway salesperson dressed as Mrs. Doubtfire.

While the idea of salespeople doing business inside your home might sound appalling, the bigger struggle for many people today, sadly, is a battle against alienation and isolation inside one's home. Just as porn has replaced sex, the clinical efficiency and convenience that come with technology is reducing direct human contact everywhere. There is a growing population for whom a chance to interact with a visiting human—even one that is working on commission—is a salve for loneliness. Retail therapy may soon come to the bedside in the form of an attentive service provider not unlike Mrs. Doubtfire, only more seductive.

As unseemly as that future may seem, the history of retail, looking back, has been one remarkable step-function after another of increasing trust between retailers and consumers. In the beginning, folks traded products guardedly in an open market. Then folks allowed others to enter their store and negotiate for products that were out of their reach. Then someone invented stores that allowed complete strangers to trespass onto their property and fondle undefended assets before deciding to pay the list price. Then someone invented the idea that you could return products after purchase. Then someone invented the idea of shipping you nine boxes of shoes and allowing you to return eight.

Each of these retail revolutions required a new level of trust that may have been unthinkable to prior generations. One could make the argument that each of these revolutions benefited both the retailers and consumers. But how will it play out going forward?

If retailers make a play for your "smart home" as the next frontier for the point of sale, the formerly trespassed are about to become the trespassers, and they know more about you than you know about yourself. One wonders if the continued encroachment will have ill consequences.

The most vulnerable to this in-home sales model are the elderly. Even if they have money, they are starved for attention and often not able to get out of the house on their own. The elderly in America are largely left alone by their families, and their everyday human contact is more often with service providers such as home-care personnel and nurses. Under these circumstances, the trust that the elderly confer on these providers tends to be high—a home-care provider might spend thirty hours in the house per week—and is subject to the risk of abuse. Fortunately, outright abuses such as inheritance embezzlement are rare. However, a vision of a future where in-home service providers motivated by sales commission are influencing the purchasing behavior of the vulnerable is one that should give all of us pause.

Ultimately, this is not a story about any one particular home delivery service. The invisible hand is a soulless race to the bottom or, more accurately, to the bottom line. If one company doesn't backdoor future salespeople through your front door masquerading as service providers, another company will gladly step in.

We may be about to open the door to that future.

Intertemporal Hegemony

Editor's Note: This frame also applies to intertemporal imperialism and colonialism.

Whether under the epistemic guise of imperialism, colonialism, or mercantilism, the polemic of hegemony has always been the same: the empowered preferentially benefit while subordinates disproportionately bear the costs. While this story has been told in three-dimensional space—the Ottomans versus their vassal states or the British Empire versus nearly everyone under the never-setting sun—the time has come to recognize its occurrence in four dimensions: intertemporal hegemony.

The ontogenesis of intertemporal hegemony can be framed in the following way. At its roots, hegemony is enforced by violence. But since the fitness cost of violence is high, multilevel selection favors the emergence of less lethal proxy behaviors (e.g., pecking behavior among chickens to establish rank). Indeed, human social evolution appears to progress toward less deadly and more accommodative forms of hegemony: actual violence codified as implied violence and implied violence codified as social contracts, reinforced by the powerful institutions that enforce them. Along the way, as subordinate groups are incorporated into these institutions—hegemony masquerading as husbandry—the awareness of the hazing can disappear into the haze.

From the combinatorial capacity of social contract innovations emerged perhaps its most insidious mutation: the intertemporal contract. Like a time machine, this novel type of contract enables a hegemon to manifest its destiny over four dimensions by rolling its metaphorical tanks over the future. That is to say, groups living in one time period can extract value from yet unborn groups while saddling the latter with the costs.

Take, for example, pension funds. In theory, the intent is appealing: a cohort of pensioners pool their savings into an investment vehicle that pays out later, during their retirement years. However, since pension funds pool contributions from cohorts of different vintages, a risk of wealth transfer among these cohorts arises. Critics have called the model—using the investment contribution of a later cohort to pay back an earlier one—by less flattering names. Likewise, Social Security, a public version of this practice, amounts to a social contract between generations that appears increasingly doomed to leave lesser amounts to those who haven't yet signed the obligatory contract.

The federal deficit is another money tube of intergenerational wealth transfer. When a group that has no voice or vote in the matter—such as yet-unborn citizens—is forced to repay a debt that another group owes, that's not "borrowing from the future."[112] That's confiscation from the future. Such taxation without representation is the precise reason the United States declared independence from the British Crown in the first place.

Once seen once, the watermark of intertemporal hegemony becomes visible everywhere. When young medical interns cleaning patient latrines are told that their discounted wages are a way of paying dues that they will harvest later in their careers, they are not being told the truth. In reality, incumbent elites are extracting labor contributions from interns while dumping the dirty work on them. When the interns become the incumbent elites, they perpetuate the vicious cycle of intertemporal hegemony with the next generation. Similarly, entry-level staff on Wall Street, Main Street, and in graduate schools serve their vassal lords in a self-perpetuating cycle of intergenerational extraction.

Indeed, much of the current condition of human sociality reflects the consequences of prior intertemporal hegemony. Each new generation of young people is chiseled by "the system" into becoming line workers among various institutions in the industrial complex whose very efforts unconsciously perpetuate the system's hegemony in a self-reinforcing fashion over time. The fate of youth today has already been pre-mined and incorporated into the net present value of existing stakeholders through the portal of intertemporal social contracts.

Seen through a wider lens, intertemporal hegemony is an epiphenomenon of an evolutionary dislocation: kin skin in the game failed to scale as the operating

112. See http://www.crfb.org/blogs/budget-projections-debt-will-exceed-size-economy-year.

system of human eusociality as we globalized, which resulted in the emergence of exclusive stakeholding as the dominant modality of intergroup interactions. The large-scale ills of human history—nepotism, despotism, cronyism, nativism, nationalism, mercantilism, colonialism, and imperialism—all relate to the abuse of excluded stakeholders. Intertemporal hegemony is just a new version of the same old game played across time.

In the tree of life, it is becoming increasingly apparent that humanity has found itself out on a limb. To some extent, intertemporal competition is an inescapable part of evolution—the earlier birds that get the worm deplete the commons—but the emergence of intertemporal contracts has enabled earlier birds to colonize the future through the wormhole of time. Whereas systems that regenerate more value for the future are favored by selection over the long haul, one that cannibalizes the future—such as human civilization today—faces existential risk.

If there is a saving grace, it is that humanity is perhaps the first species to consciously care about leaving a better planet. But, it also needs a workable plan. Updating humanity's bioalgorithms of inclusive fitness with social algorithms of inclusive stakeholding, including future stakeholders, is our greatest existential hope for sustainable prosperity.

Moral Imperialism

On July 16, 2019, the founder of the space company Blue Origin outlined a plan to preserve a dying Earth by "using the resources of space" and moving high-polluting industries like manufacturing to other planets.[113] The deeper message may have flown over everyone's heads.

The strategy of exporting problems to distant lands while extracting resources from them is nothing new; the idea is as old as the history of imperialism, and space is merely its next frontier.

As the expanding concentric circles of imperialism have hit their planetary limits, we can see the accumulated consequences: pollution, consumerism, and dystopic desperation. Growing by extracting resources from others is the self-defeating algorithm of cancer that eventually kills the host. Interplanetary imperialism merely kicks the unrecycled can down the road.

But hold your revolutionary high horses. We too are blind to our own everyday imperialism.

Participation is our imperial complicity. In the illusory emotional safety of morning coffees and pajamas, our digital keystrokes extract natural resources from anonymous mines in other continents to deliver disposable goods that become permanent trash shipped to anywhere but our own backyard. It is hard to rebel against digital imperialism while accepting the trickle-down loot garnered through it.

But people don't call themselves imperialists. The word is only used to label the behavior of others, even though we are each other's "other."

Which brings us to a larger point. Even our protestations reflect moral imperialism: extract the credit and dump the blame on others.

At the end of the day, everyone participates in the greater system in a way that may cause some harm to others, however subtle or remote, but there is a general

113. See https://www.cbsnews.com/news/jeff-bezos-caroline-kennedy-apollo-11-future-space-exploration-2019-07-16/.

blindness to this reality. Our minds tend to downplay our role in the harm we cause others and overplay the role of others in the harm we experience.

Self-expanding, race-to-the-bottom beasts are everywhere—in the food we eat, the social networks we track, the pills we take, the sports teams we follow, the political machines we support, and the self-help groups we join. What we miss—nearly all of us, nearly all of the time—is something more fundamental: that all of these self-dealing, race-to-the-bottom organizations have grown out of *the same global petri dish of misaligned incentives* that is modernity. And all of them feed on the same set of primal instincts that allowed us to survive prehistory.

What's notable is that we have little trouble seeing others as evil. We easily recognize them and call them out in the institutions of which we are not a part. Corporate executives and entrepreneurs have no difficulty recognizing the self-dealing superorganism that is government bureaucracy; noble civil servants (a.k.a. government bureaucrats) have similarly little difficulty recognizing the self-dealing beasts on Wall Street and in Silicon Valley. Pulitzer Prize-winning news organizations understand that their duty is to reveal the self-dealing beast that is the political system. Politicians assert that it is their duty to reveal the fundamental corruption of the media.

The general blindness to this reality is evidenced by Goodwin's Law, the Internet rule asserting that, "as an online discussion grows longer, the probability of a comparison involving Nazis or Hitler approaches 1." Importantly, this law holds regardless of the characteristics or the views of the individuals engaged in such discussions. It evidences the fact that our society has evolved (or cultivated) a deep-seated need to experience not only as evil but as absolute evil all those who oppose our many and varied heroic quests.

We all gaze across the moats that divide us and point to the self-dealing beasts on the other side. We Occupy. We Tea Party. We Yellow Jacket. We sue. We counter-sue and counter-counter-sue. We Tweet, we post. Above all, we attack. We do so fueled by the conviction that the wind of history is behind us.

Yet, even if we were to prevail in demolishing all the beasts we perceive around us, we would be seen as the beast by others who want to take us down.

And so it is that we proceed on seven billion hero's journeys that add up to a single collective ride on the Carousel of History.

Dissociality Supercycle

Editor's Note: Ultimately, dissocial civilizations die unless they grow. Since growth eventually hits limits, the sentence can be shortened to: dissocial civilizations die.

In dissocial civilizations, power structures are contrivances: hard to build and inherently unstable. They are transiently propped up through growth that keeps mercenaries loyal. When growth ends, which is inevitable in a finite resource world, the system of mercenaries collapses quickly in a chain reaction back to self-stable, underlying components—kin-based autonomous eusocial units. Thereafter, social entropy reboots the journey from decentralized eusocial hives to centralized dissocial civilizations.

This historical recursion is hereby referred to as the dissociality supercycle.

Dissocial civilizations are characterized by cultural advancements: feats of engineering, science, and art are the spoils of intense competition and centralized power. For example, the emergence of numeric and linguistic systems during dissocial eras may make later historians swoon, but they represent symptoms of mutual low trust: numeric and linguistic systems arise for the administrative purposes of counting the spoils and accounting for the events. Science and engineering may originate for their own purposes, but their widespread adoption and co-option tends to strengthen existing power structures. Some even argue that what was later classified as the art of dissocial civilizations was often symbolic propaganda commissioned and underwritten by power structures to exert command.[114]

These eras of hegemony by power structures, dehumanization, and prolific cultural output are widely referred to by historians as periods of "enlightenment."

114. See https://scholarship.tricolib.brynmawr.edu/bitstream/handle/10066/714/2003GlennM.pdf?sequence=5#:~:text=Architecture%20is%20still%20intricately%20tied,prominence%20has%20become%20less%20explicit.&text=Architecture%20therefore%20demonstrates%20power%20through,his%20power%20as%20a%20leader.

No doubt, one could make a case that the output of high culture from these eras has contributed immensely to progress in the long run. In the immediate, however, that high culture is also often an epiphenomenon of the leisure surplus harvested by the aristocracy through exclusive stakeholding and exploitation of the masses. Of note: many of America's founding fathers owned slaves.[115]

With all that as background, the disintegration dynamics of dissocial civilizations are worthy of closer inspection. The rapidity of decimation among interconnected power structures speaks to their frailty.

One of these hyperdynamic periods is the collapse of the Late Bronze Age.

The Mycenaeans, Kassites, Hittites, Ugarits, Amorites, Luwians, and Canaans are among the cohort of gilded dissocial civilizations that had taken centuries to build but dissolved during the same 50-year span from 1200 BC to 1150 BC. Once the disintegration began, almost all major power centers in these regions, including virtually every magnificent city, were swiftly destroyed and abandoned. Many of them would never be occupied again, including Hattusa, Mycenae, and Ugarit.

A range of explanations for the systemic collapse has been posited by historians, including the usual suspects: climatic events, pandemics, external invasions by groups such as the Sea Peoples, and the failure of political, social, and economic systems.[116]

A common factor underlying these disparate theories, however, is the lack of underlying resilience of dissocial systems. To some extent, the increasing complexity and specialization of the Late Bronze Age political, economic, and social organizations in these civilizations can be seen as protecting existing power structures.[117] Yet, their inability to sustain some level of organization in the face of trauma or changing conditions bespeaks the inherent fragility of dissocial civilizations.

115. See https://www.smithsonianmag.com/history/founding-fathers-and-slaveholders-72262393/.
116. See https://www.ncbi.nlm.nih.gov/pmc/articles/PMC7123324/.
117. Thomas, C. G., & Conant, C. (1999). *Citadel to City-State: The Transformation of Greece, 1200–700 B.C.E.* Bloomington, IN: Indiana University Press.

Indeed, scholars such as Carol Thomas, Eric Cline, and Craig Conant have argued that the fatal flaws of the Late Bronze Age were centralization, specialization, complexity, and top-heavy political structure—weaknesses that were exposed by stochastic events (examples: volcanic eruptions, peasant revolt, drought, crop failure, famine, mercenary defections, overpopulation, invasions by Sea Peoples and Dorians).[118] In the opinion of these scholars, the inherent brittleness could explain why the collapse was so widespread and able to render the Late Bronze Age civilizations incapable of recovery.[119]

But one can make the following axiomatic distinction.

One can argue that the holistic collapse of the Late Bronze Age is evidence of the intrinsic fragility of dissociality, and not of civilization itself. To wit, the downward spiral of the collapse of centralized dissocial power structures could only have ground to a halt once it had sufficiently atomized into an autonomous collection of kin communities—in other words, into their self-stable eusocial components. It's a bit like the radioactive instability of uranium decomposing into more self-stable levels of atomic organization such as lead.

Here is the bigger picture. As it stands today, the envelope of human possibilities is constrained between two goal posts: races to the middle (high culture), whose emergent power structures eventually self-implode due to intrinsic instability, and races to the bottom (the actual periods of collapse), where the final bottom represents fragmentation into decentralized self-stable eusocial hives of low cultural output (i.e., not much decorative pottery).

Looking ahead, the only way we can achieve sufficient escape velocity as a civilization—to venture beyond the neverending yin-yang toggling between races to the bottom, where heads roll, and races to the middle, where eyes roll—is to instantiate social algorithms of inclusive stakeholding. Then, and only then, can we step off the Syssiphian treadmill of recursive devolution back into a collection

118. Cline, E. H. (2014). *1177 B.C.: The Year Civilization Collapsed.* Princeton, NJ: Princeton University Press.
119. See https://en.wikipedia.org/wiki/Late_Bronze_Age_collapse.

of decentralized feuding eusocial hives (or their more malignant cousins, the feudal hives). Thereafter, humanity will finally be able to embark on a race to the top toward a global eusocial hive.

Historic Recursion

It is often said that history repeats itself. Others have argued that it rhymes, rather than repeats. Still others changed the tune a bit to make essentially the same point.

What they are referring to is sometimes called historic recursion, the seemingly endless "Groundhog's Day" that is history: the rise and fall of empires, oscillations of polity, marches toward war, etc.

The idea echoes the philosophy of Eternal Return, that everything in the universe recurred and will recur in a self-similar fashion an infinite number of times across infinite time or space. Variations on this philosophy can be seen in schools of thought across time and space: Indian antiquity, Judaic literature, Friedrich Nietzsche, etc. The philosophy recursively embodies itself.

To some extent, the *deja vu* feeling of history is understandable. It seems like there's always someone taking the hill, someone crossing the Rubicon, and someone turning a blind eye.

But the larger blindness has been our inability to *see past* history—in order to see an even larger frame.

In the beginning, nature embodied eusociality as kin skin in the game—the biological version of inclusive stakeholding. After the Promethean Fire, humanity outgrew kin eusociality and formed dissocial civilizations without updating the code of inclusive stakeholding.

As a result, there always have been excluded stakeholders who pay the price of a blinded eye—the price of tyranny, slavery, colonialism, racism, sexism, elitism, pollution, extractive capitalism, corruption, and inequality.

Today, beyond feeding our own, humanity continues to grapple with how to live together in an interconnected global melting pot of self-interests. While conflagrations still flare and debates about merit and redistribution rage, the good news is that, in many modern societies, prosocial intentions and policies have moderated aggressions and leaned into equity and justice.

Yet, if there is anything to learn from history, it is that prosocial intentions are eventually undermined by self-interest, the very force that necessitated them in the first place. In the long arc of history, seemingly good governance or progressive economic systems tend to be self-consumed by races to the bottom among self-serving power structures, eventually precipitating collapse or revolt.

To break orbit from this Carousel of History is to see that historic recursion is a symptom of our repeating failure to update the bioalgorithms of eusociality with dissocial ones. Rather than capitulating to the recapitulations of history, retreading the same old ground wondering why it's *deja vu* all over again, it is time to look up and see that the arc of history bends toward inclusive stakeholding.

Are Dark Ages and Enlightenments Reversed?

After the Late Bronze Age collapse, Mycenaean civilization dissolved into the so-called Greek Dark Ages, when society atomized into a decentralized collective of eusocial kin hives. Archaeologists have noted that a striking feature of this era is the lack of material cultural output: People stopped painting and building palatial centers.[120] [121] As trade routes vanished and kin-based villages reorganized themselves to be self-sufficient, writing in the Linear B script ceased.[122] The end of redistributive economies meant there was no longer a need to keep records on commerce.[123] The world of administrative officials disappeared.[124]

The little material output from this era that has been found has been said to lack decorative style and aesthetic cohesion, so much so that historians consider the eventual emergence of the geometric style characterized by lines and curves a telling sign of cultural revival.[125]

120. If there was a return of art to immediacy and transience—a return to oral and dance rituals innate to all social species—then such art may not have been sufficiently captured in ancient archaeological records the way such art was captured in the works of the minstrels and troubadours of the later European Dark Ages. In other words, art would have returned to its prehistoric forms.
121. See https://en.wikipedia.org/wiki/Greek_Dark_Ages.
122. Stanton, A. L., Ramsay, E., Seybolt, P. J., & Elliott, C. M. (Eds.). (2012). *Cultural Sociology of the Middle East, Asia, & Africa: An Encyclopedia.* Thousand Oaks, CA: SAGE.
123. See http://www.perseus.tufts.edu/hopper/text?doc=Perseus%3Atext%3A1999.04.0009%3Achapter%3D3.
124. See https://en.wikipedia.org/wiki/Greek_Dark_Ages.
125. Whitley, "Social Diversity," pp. 342, 344ff. Also see https://en.wikipedia.org/wiki/Geometric_art. Archaeology suggests that the advanced civilization of Mycenae began to collapse around 1200 BC. By 1050 BC, the recognizable features of Mycenaean culture were gone. What replaced the centralized culture were decentralized clusters of eusocial settlements: kin hives, or *oikoi*. For example, excavations of Nichoria in the Peloponnese reveals a Bronze Age town that was abandoned in 1150 BC but repopulated as a small village cluster by 1075 BC. It is thought that forty families lived there as a self-sufficient farming community. The remains of a 10th-century building presumed to be the chieftain's house suggested that it was made from the same materials (mud brick and thatched roof) as the rest of the village and reflected a standard of living not much different than others of their village—an indicator of eusociality. It is likely there were some settlements that retained dissocial features during this predominantly eusocial era. Lefkandi on the island of Euboea appears to have recovered quickly after the collapse of Mycenae. In fact, in 1981, excavators of a burial ground found the largest 10th-century building yet known in Greece; referred to as the *heroon,* the structure is renowned for its showy display of wealth, status, and access to trade. That the settlement soon disappeared has been interpreted by James Whitley as a sign that Lefkandi had a "Big Man" social structure—dissocial and therefore inherently unstable. The formal end of this post-Mycenae era of eusociality in Greece—the resumption of increased social entropy, relationship liquidity, and diasporas—had probably

That said, the view that the Greek Dark Ages lacked culture because it lacked material production, decorative style, and artistic cohesion is just one perspective. One might alternatively argue that the return of decoration and cohesive style coincides with the return to a less authentic, more transactional culture in which artifacts serve purposes beyond their utilitarian function: for example, self-promotion or commercial success.

Indeed, the entire convention of referring to the post-Late Bronze Age eusocial era—when the need for numbers and words faded and the pottery left behind is less decorated and less common—as the Greek Dark Ages begs larger questions. The imagery of dark and light periods of history stems in part from our lack of knowledge about these post-collapse periods, due to the aforementioned reduction of cultural output.[126]

To be sure, one could strongly argue that periods of Enlightenment have been associated with high cultural output and material progress that brought many people out of poverty and subsistence existence. But the pejorative meaning of the word "dark" to describe pockets of eusociality and low cultural output—lesser known epochs between peak civilizations, indigenous cultures, prehistoric populations—may reflect a conflation, perhaps even a projection, of our ignorance of these people as *their* ignorance.[127] This cognitive dissonance is evident even among well-intended artists who themselves attempt to shine the light on the foibles of our nature, such as Joseph Conrad's purported blindness to his own racism in *Heart of Darkness.*[128]

already occurred by the beginning of the 8th century BC. By this point, cemeteries, sanctuaries, and colossal free-standing temples were richly adorned with exotic items from Near Egypt, including amber and ivory, indicating that commercial transaction—and commodification—had begun by this time. The decoration of pottery became more elaborate, suggesting a return to the performative era of human interactions. The return of governance by elite aristocrats rather than by the single *basileus* or chieftain of earlier periods suggested the return of dissociality. And, just like that, the brief respite of eusociality was over. Unlike during prehistoric times—before the Promethean Fire—humanity's retained ability to harness energy, which promotes social entropy, relationship liquidity, diasporas, and a transactional culture, ensures that eusocial periods don't last very long before dissociality returns again.

126. Thompson, B. (1996). *Humanists and Reformers: A History of the Renaissance and Reformation.* Grand Rapids, MI: Erdmans, p. 13.

127. At least based on the length of their Wikipedia entries, it seems that we know quite a bit about knowledge and are ignorant about ignorance; https://en.wikipedia.org/wiki/Ignorance https://en.wikipedia.org/wiki/Knowledge.

128. See https://disa.ukzn.ac.za/sites/default/files/DC%20Metadata%20Files/Centre%20for%20African%20Literary%20Studies/ALS%204_4_6_1_21/ALS%204_4_6_1_21.pdf.

Indeed, denigrating the regional diversity of material production during the Greek Dark Ages as "lacking artistic cohesion" could be interpreted as gatekeeping by arthouse power structures suffering from superiority complexes associated with intellectual imperialism.[129] If anything, one could argue that failing to see the beauty of a regional diversity of material production during eusocial eras reveals more about the darkness of the art world during dissocial epochs than about the art of the Dark Ages.[130] It could alternatively be said that periods of aesthetic cohesion reflecting gatekeeping, conformity, and performative inauthenticity among commercially motivated and status-promoting artists are the true Dark Ages, masquerading as the Enlightened Age.[131] Some might even argue that one of those periods is now.

Looking back now, could the opposite case be made on the naming convention? That is, should historians reclassify the eusocial "dark" eras that produce a diversity of authentic artifacts as "enlightened"? Should they also rename the dissocial eras that produce convergent commodified artifacts in commercially motivated markets—sometimes referred to as "sellouts"—as the Dark Ages?

The answer here is far from black and white.

No doubt, the hidden and often forgotten price paid by large swaths of genuine art and artists excluded from art power structures during "go-go" dissocial eras are not insignificant. From the perspective of those kept in the shadows by dissocial power structures, the so-called enlightened eras may have felt like the darkest of periods. Who knows what else the world missed from Herman Melville, Johann Sebastian Bach, Vincent van Gogh, Zora Neale Hurston, Georges Méliès, Louis Sullivan—let alone the unpursued works and careers of other talented upstarts—due to systemic oppression of outlier art by the gatekeepers of their eras?[132] The benefits of Copernicus's work might be considered a marvel enabled

129. "The most striking feature of the Dark Ages is its regionalism, its material diversity." Whitley, J. (1991). "Social Diversity in Dark Age Greece." *The Annual of the British School at Athens* (86): 342, 344.
130. Whitley, "Social Diversity," pp. 342, 344ff.
131. See https://www.theatlantic.com/technology/archive/2020/08/why-every-city-feels-same-now/615556/.
132. See https://www.britannica.com/list/6-rediscovered-artists.

by the printing press bringing the end of the Dark Ages, but oppressive gatekeeping by dissocial power structures of the Enlightened—domestic intellectual imperialism—kept the world in the dark about Copernicus's theory nonetheless.

All that said, many today would argue that renaming the Enlightenment Ages the Dark Ages would be—like some of Melville's best-known characters—going a bit overboard. Dissocial eras, despite their untoward tendencies described above, have contributed enormously to humanity's positive transformation. Even our knowledge about the art of eusocial eras has been enabled by some of the very progress made during dissocial eras that put us in a position to contemplate ourselves more fully.

Either way, the benefits of rebranding the Dark Ages for their positive features could put us in a position to make the best of both worlds. Imagine fusing the social cohesion of eusociality with the productive cultural output of a civilization driven by competition and self-interest—language, math, knowledge, art, and other byproducts of feeling a bit under siege from counterparty risk.

A society that can combine the best of these features—clicks without bait, telephone calls without telemarketers, nuclear energy without nuclear bombs, etc.—may one day be remembered as a truly enlightened society.

CHAPTER FIVE

Fifth Order: Race to the Middle

Reader's Guide to Chapter Five

Optimists claim the tide of the human condition has been rising. Indeed, competing self-interests as well as prosocial attitudes have brightened human life: increasing life spans, the decline of physical combats, etc.

But shadows also lurk. The gains have often been paid with the suffering of excluded stakeholders. Moreover, assessments about the human condition depend on what temporal scale is employed. As with tides, zooming in or out reveal smaller and larger oscillations—the ups and downs of history.

Chapter Five argues that dissocial power structures constrain historical tides to oscillations between a race to the bottom and a "Race to the Middle," as detailed in the eponymous first essay.

A race to the bottom puts elites at risk of a coup, motivating power structures to evolve instead towards more accommodative forms of hegemony—ones based on free will and free markets. This race to the middle leads systems to stall. The competition to nowhere exhibits bureaucratic dynamics discussed in "Byzantine Reflex."

However, don't confuse the bureaucratic stasis as a sign that it is necessarily passive. It's quite often passive-aggressive. Besides their clinical efficiency in disorienting potential dissidents with their inefficiency, race-to-the-middle beasts also subvert them by enlisting their services. "Neutron Bomb of Human Sociality" describes how institutions pit people against each other, undermining their ability to combine forces against the institutions.

As races to the middle put people to sleep with the lullaby of cultural blandness, some may start running screaming from, or possibly into, the abyss. That's one way to describe Burning Man, as characterized in "Counterculture."

It is thus that all roads may lead to Rome in races to the bottom, but in races to the middle, all roads lead to nowhere. That leaves us with a dawning realization that, based on the current arc of history, "The Closer We Get, the Farther We Are" from humanity's best destiny.

Race to the Middle

Editor's Note: Whereas races to the bottom inspire coups, races to the middle produce a tranquilizing elixir, a blend of blandness and mediocrity that dulls the senses.

Is the race to the bottom bottomless? Aren't races supposed to end?

Maybe.

If a race to the bottom becomes sufficiently egregious, a revolt by the exploited might overthrow the establishment. At least that is a view often served up as a temporary salve, however fleeting, for the plague of races to the bottom everywhere.

The possibility of revolt, including the kind in which heads end up on the chopping block, is evident to those in power. Thus, to some extent, the oppressors have a vested interest in putting the brakes on a runaway race to the bottom—not because they have a sense of justice but to save their own necks. Even in this situation, the power of vested self-interest is hard at work.

Concessions by those with political power—including term limits and a system of checks and balances—put the brakes on the self-reinforcing, race-to-the-bottom nature of that power.[133] For those with economic power, concessions—including progressive taxation, philanthropy, and virtue signaling—put the brakes on the self-reinforcing, race-to-the-bottom nature of wealth inequality.[134] If a revolt becomes necessary, some of the empowered may take the side of the revolutionaries, for a wide variety of claimed and actual reasons.

What this means is that many races to the bottom eventually correct course to become races to the middle. In some ways, a race to the middle is worse than a race

133. Also known as a system of checks written to the account balances of political institutions. Yun, J., Yun, J., & Yun, C. (2019, February 15). *Interdependent Capitalism: Redesigning the Social Contract through Inclusive Stakeholding* (Kindle edition); https://www.amazon.com/Inclusive-Stakeholding-Revolution-Social-Contract-ebook/dp/B07DP6LPPV.

134. See https://en.wikipedia.org/wiki/Virtue_signalling.

to the bottom, as it deters the cleansing of institutions of their latent extractions. On the other hand, races to the middle throughout history have generally helped guide the trajectory of human life upwards, thereby lifting all boats. In short, the race to the middle is a curse and a gift.

The race to the middle is one of the tools the beast uses to maintain the status quo. It provides a salve for second-order symptoms—but not the first-order root cause—of issues, which can temporarily temper the risk of rebellion. As discussed previously, the root cause of why competition selects for power rather than efficiency is exclusionary stakeholding—or the lack of inclusive stakeholding. Once the power structures are in place, the perpetuation of hegemony through the domestication of revolutionary ardor is the beast's objective.

Here are some examples.

Many business, social, and political leaders today have cited humankind's unbelievable material progress—the number of PhDs, toilets, easy access to food and clean water, technological innovation—as indicators that life is far better today than ever before. Those claims are unassailable. The associated cultural commodification, however, has produced a dispirited population that is struggling with an increasing sense of alienation, loneliness, and lack of meaning in their lives. This combination of spiritual regression and material progress lands us approximately in the local minima that is the race to the middle, where the blended average of quality of life is just good enough to deter revolt.

Other examples of the race to the middle are race and gender relations in America. Every well-intended sequential concession by the status quo has domesticated the discontent without fully addressing the underlying issues of non-inclusive stakeholding. The mixed blessing of the race to the middle can leave entire populations in no-man's land or, in the case of the suzerainty offered to the Native Americans, in a marginally habitable habitat. Whereas the race to the bottom inspired Tecumseh to fight for justice, the race to the middle inspires fights for casinos.

Universal basic income is another example. It's a post-hoc salve that perpetuates a system that invariably self-generates wealth inequality—a fiat bribe of the public that distracts from fixing the underlying cause: exclusionary stakeholding. Again, treating the symptoms merely kicks the can—or the can'ts—down the road.

On the surface, universal basic income might look superficially similar to our proposal: universal basic stakeholding. In both cases, the accrual appears tethered to the success of the nation. There is, however, a small and important distinction: whereas stakeholding, as in universal basic stakeholding, allows a direct contractual connection between the dividend and aggregate success of the nation, the connection in universal basic income is seen as indirect and subject to distortion.[135] Similarly, if public and private pension funds were organized around stakes rather than defined payouts, conflict between stakeholders would be reduced and alignment increased.

The race-to-the-middle process is rapidly tranquilizing many of our institutions to a standstill of mediocrity: education, science, medicine, food, media, politics, art—you name it. It's not just, "Where have you gone, Joe DiMaggio?" It's more like, "Where the hell is everybody?"

And that is the larger point. The most significant cost of the race to the middle is the opportunity cost of not setting up a race to the top of collective greatness. Whereas competition with alignment creates a race for the top, malalignment with competition surfaces power that drives the system first toward the bottom and eventually toward some blasé equilibrium in the middle.

The race to mediocrity produces a sense of cultural blandness and banality that is not bad enough or good enough to get too excited about. Think about why the corpocracy of automobile manufacturing inevitably not only drives itself to mass produce Toyota Camrys but induces the entire industry to make approximate replicas of the Camry. It is an excellent car, but it also is stupefyingly ordinary, relative to what is possible for the times. The inherent self-dealing culture of middle management, which lacks proper stakeholding at every level of the chain, ensures that imagination gets winnowed out in service to a biweekly direct deposit.

135. This difference in perception is similar to the distinction between when an employee earns a dividend as a company stockholder versus earning a year-end bonus. In the former case, the binding nature of the dividend that comes with being a stakeholder creates a different set of motivations than a year-end performance bonus that is subject to many forces beyond appreciation of the stock's value. Universal basic income has the potential to domesticate public resentment against self-serving institutions.

Through commodification, the race to the middle has infected almost every vector of human life, where the feckless total output of the industrial complex is the Toyota Camry of everything—tranquilizing yet tranquil, venal yet banal. A lullaby of domesticated bliss.

Byzantine Reflex

Once an organization or an institution gains power, it has a natural incentive to protect that power and maintain the status quo.[136] In medieval times, such protection might have taken the form of a castle topped with battlements and surrounded by a moat. Modern-day institutions not only employ updated versions of a moat but use a subtler approach to protect themselves, like the one Dame Gothel used to hide Rapunzel in the Grimm fairy tale: a self-cultivated thicket.

Industrial economists understand well the tendency of powerful incumbents to use their influence on governments to construct elaborate "regulatory thickets" that inhibit competitive challenge. Similarly, in monarchical eras, courtesans constructed similarly elaborate, almost impenetrable rituals to protect their positions of influence and power. The term "Byzantine complexity" refers to the impossibly complicated protocols and procedures of the Byzantine Empire, which was the continuation of the Roman Empire in its eastern provinces in late antiquity and the Middle Ages.[137] The infamous rituals of the Byzantine court, however, were exceptionally complex only in comparison with the norms of the era, when most monarchs were elevated feudal lords and royal courts were relatively rough-hewn affairs.

While most of the complexity that allows modern exchange economies to exist is deeply submerged in digital protocols and, increasingly, nearly inscrutable machine learning algorithms, enough remains on the surface to make the Byzantine court look like a family affair. This is not because complexity tends to increase within any institution over time (as was largely true of the Byzantine

136. This is also true within the "bureaucracy of the mind" that lives inside each of our brains.
137. "Guests at royal banquets were assigned titles that denoted where they could sit in relation to the emperor, whom they could talk to, and what they were allowed to discuss. Eventually, the rituals became so complex that treatises were written to help outsiders understand proper etiquette, and the emperor employed officials to teach newbies how to behave." See Palmer, B. (2011, October 20). "How Complicated Was the Byzantine Empire?" *Slate;* https://slate.com/news-and-politics/2011/10/the-byzantine-tax-code-how-complicated-was-byzantium-anyway.html.

Empire) but because powerful incumbents have learned to use complexity as a barrier against competition.

The Byzantine reflex refers to systems that are characterized by a tendency to self-evolve toward greater complexity and obfuscation in a way that favors the asymmetric beneficiaries of the status quo.[138] Indeed, such systems will self-select against simplicity and transparency. A corrupt government, for example, will do nothing to stop complicating government regulations, as such obfuscation enables them to set predatory traps for the public.

At one extreme, power structures may eliminate choice altogether. In a rational healthcare system, for example, patients ought to be able to choose their physician and the physician ought to be able to choose the right treatment for the patient. Instead, the "beastlike" healthcare system has self-organized and consolidated its power to the point where patients often can't pick their doctors and doctors often can't choose the treatments that might be best for their patients.

The larger point is this: unless interests are properly aligned through inclusive stakeholding, the natural inclinations of individuals and free markets will naturally self-assemble into the exact nightmare that libertarians fear: a bureaucracy.

This is how America, a nation built on the libertarian principles of individual freedoms (a.k.a mutually exclusive stakeholding), evolved into a restrictive legal system that constrains individual freedoms. Today, the number of laws in the nation founded on seven articles of the Constitution has been continually self-expanding, to the point where the most accurate answer to the question "How many statutes are there in the U.S. government?" is that it is unknowable—according to the U.S. government.[139]

138. One example of this phenomenon is the evolution of high-potency problems to low-potency ones. Eusocial groups are biomotivated to respond to issues large and small commensurately, but that is not necessarily so in dissocial groups. On the one hand, high potency threats are highly problematic; however, they also tend to get identified and addressed. Meanwhile, low-potency threats pose a different kind of challenge in ESH societies. More broadly, the race to the middle selects for low-grade problems of every variety, the effects of which can accumulate throughout the system, until it becomes the system. It is thus that in an ESH world, low-potency threats may prove to be more persistent and ultimately more toxic than high-potency ones. These are the types of phenomena that leave people with an uneasy feeling about the world: that something is a bit off, but they can't quite put their finger on why.

139. Cali, J. (2013, March 12). "Response to 'Frequent Reference Question: How Many

Here is the take-home message: without aligned interests to protect you from abuses in any system of Byzantine complexity, be wary of an underlying counter-party that could be acting against your interests.

Federal Laws Are There?'" (Library of Congress post); https://blogs.loc.gov/law/2013/03/frequent-reference-question-how-many-federal-laws-are-there/.

Neutron Bomb of Human Sociality

The fear of potential mutual nuclear annihilation during the Cold War was subverted by institutions to domesticate individuality. In a broader sense, exclusive stakeholding is the invisible neutron bomb of human sociality that has already been detonated and is radiating the cancer of dissocial institutions across the globe.

In his novel *One Flew Over the Cuckoo's Nest,* Ken Kesey explores the power institutions wielded over individuals during the Cold War era through the subversion of innate human instincts. That power is nested in Nurse Ratched, who holds dominion over her patients in the psychiatric hospital. She manifests that power in part by leveraging the inmates' self-dealing instincts: she baits them to disclose damaging information about each other in exchange for her granting favorable wake-up times.

She herself is a tool, an agent of a larger authoritarian system that maintains authority by undermining its subjects' ability to revolt against the system. In that sense, even the Cold War itself was an example of two superorganisms—America and the USSR—that had been domesticated by an even larger system-wide superorganism that kept each nation fighting against each other and domesticating their own people by pitting individuals against each other.

A fractal recursion is evident: a global superorganism was domesticating Cold War superpowers by inducing them to fight; each Cold War superpower was domesticating its satellite states by inducing them to fight internally; each state was domesticating its citizens by inducing them to infight; a veterans mental hospital was domesticating war victims by inducing them to infight. In essence, the global superorganism maintains power by atomizing its subjects.[140]

The central irony is that Kesey depicts the system, not the inmates, as "cuckoo." In using subversion to undermine Nurse Ratched's subversion, McMurphy turns

140. The self-expanding beast does the same thing to science, atomizing systemic concepts into its inscrutable individual components.

the cuckoo's game on its head. In nature, the cuckoo preys on other birds' nests, but in Kesey's novel, it's the free-wheeling geese that fly over the cuckoo's own nest. The narrative is part of Kesey's own revolt against what he sees as the broader tyranny of American institutions in the 1950s: tranquilization marketed as tranquility and domestication marketed as domestic bliss. Yet, if you wind the cuckoo clock five dozen years forward, we've circled back to where we started. Institutions everywhere are competing to control individual minds, witch hunts are polarizing the nation, and modern-day "Merry Pranksters" are running to the hills with psychedelics.

It's crazy.

Counterculture

Editor's Note: Dissociality promotes cultural commodification. Counterculture follows.

"Welcome home." What a strange thing to say to someone you've just met, I thought. Plus, this was not my idea of home. This was actually just about the opposite of my idea of home—the shelter I had left behind eight hours earlier that had suburban comforts such as copper plumbing and air conditioning. It was the end of summer, and I was on my way to attend my first Burning Man, an annual performance art of 70,000 volunteer refugees hauling an entire Walmart Supercenter to an evaporated lake bed.

Like most normal people, of course, I had resisted years of beckoning by people I otherwise consider friends to come to this blighted place—a Habitat *Not* for Humanity. The more pictures they showed and stories they told, the less interested I had become. In the photos, the place looked like Pompeii just as the dust was settling. Not central Pompeii, but a lesser known trailer park near Pompeii. The people in the photos were frozen in poses and expressions that suggested they did not have time to put on their clothes when calamity struck.

Finally, one of my friends said, "Forget all the photos and stories. If you trust me, you should join us." A beautiful paean to friendship. Or possibly how the Devil pitched Faust. I'm sure the Devil (playa name: Angel) would identify with being a "burner," and I like bargains.[141] In the broader context, a new millennium was about to dawn and I was running out of excuses. So I told them, in a tone of resignation, that I would join them. But only if I could drive my own getaway car.

At this point, I have not rolled down my window yet. There is no rush when you

141. See "playa name" defined at https://www.urbandictionary.com/define.php?term=playa%20name.

might be at the Gates of Hell.[142] Alas, I am in the middle of a dust hurricane, and—in a moment of cash register impulsiveness—I had upgraded my Ducky's car wash to include upholstery vacuuming two days earlier. The greeter tapped on the driver's side window: "Please open your trunk." That was the other reason I hadn't rolled down my window yet. If I were to describe the person I would want to welcome me home after a long trip, she was not it. Not even the lingerie part.

This particular lingerie model was sporting blue ski goggles (Walmart, $19.99). Keep in mind that I had just crossed a flat desert inferno from Wadsworth, Nevada, which is itself a desert town.[143] There was no Alpine ski hill in sight. What was really out of place, however, were the two gold cones with pointy tips protruding over her mammary glands. I decided to put on protective goggles of my own before rolling down the window.

"Please open your trunk," she requested again. Thank God, a mark of civilization. She's going to take my luggage to my room, I thought. I really hadn't given this Burning Man festival enough benefit of the doubt. I was ready to tip a lot because I had packed enough stuff to run away and start a new life in Mexico. And just what would you bring if you were starting a new life in Mexico? Certainly a lot less than what I had packed for a few days' sleepaway at Burning Man.[144]

The thought did cross my mind that she might not be the world's best partially dressed bellhop. But if not the bellhop, who was she? What did this begoggled lingerie model want? Only years later would I learn that she was a border patrol agent screening for drugs and stowaway millionaires who were leaving the American Dream in search of a refugee encampment. In my jeans and pressed shirt (Walmart, $47.06), I probably didn't look like the rest of the volunteer laborers rolling in to Burning Man nation to build semi-stable dwellings. Looking back at that customs moment, I realize we were sizing each other up.

142. If camping downwind of gases emanating from a row of 22 Port-A-Potties, which turn into fertilizer factories in the baking sun, is not Hell, then I don't know what is.
143. The town is deserted, but I repeat myself.
144. For Mexico, I would just need some cash and a sense of adventure. At Burning Man, the latter will get you far but the former won't. U.S. currency is not accepted anywhere on the playa, except to purchase three things: coffee, ice, and iced coffee.

What I best recall about that moment is that the agent wasn't wearing much except a smile, as they say, but I wouldn't have known if she was smiling because her face was covered with a bandana (Walmart, $2.99). Again, only years later would I come to understand that ambivalence is not an uncommon feature of Burning Man performance art. For the time being, however, it was hard to tell if I was about to get a car wash from the bikini-clad greeter or be carjacked and stuffed in the trunk. Anything seemed possible, and because of that, for the second time that day, prudence suggested that I turn around and drive back to San Francisco.

San Francisco was founded by Spanish colonists on June 29, 1776. Elsewhere that same week, a committee of five colonists far to the east—wearing wigs and elaborate costumes—drafted divorce papers from the establishment and gave birth to a new kind of experimental civilization. A few years later, some of them felt the need to clarify what they meant and wrote up ten basic principles to guide that new civilization.[145] But, they couldn't stop there and kept adding more to the list. Eventually, some people decided they wanted a less federal and more feral environment. So, they piled all their belongings onto cobbled-together vehicles and headed off on a dusty trail to search for a different kind of culture.

At the beginning of the 19th century, San Francisco would have qualified as a wild environment. Wild, as in there was nothing there. Then, in 1848, James Marshall announced that he had found what looked like bits of coins in the American River, and entrepreneurs from all corners suddenly upended their lives and rushed to the region to mine this particular alternative to the U.S. dollar.

San Francisco incorporated itself in 1850 to accommodate this influx of unwashed entrepreneurs. Stories from those early years suggest that it was a bit of a free-for-all: an eclectic, random mix of adventurers from all walks of life trying to build a city from scratch in a short period of time. Free-for-all, in this case, doesn't mean it was a gift economy; it meant everything possible was for

145. See https://en.wikipedia.org/wiki/United_States_Bill_of_Rights.

sale—picks, shovels, women, drugs, and dreams. To apparel this emerging class of hedonists, a cottage industry arose to make a revolutionary type of clothing the world had never seen before: the aforementioned jeans. An epicenter of Wild West mayhem was born.

A New Yorker friend once threw down the following gauntlet: between her home city and San Francisco, which had contributed more to revolutionary culture?

There is no contest. San Francisco spawned the computing revolution, the hippie revolution, the venture capital revolution, the gay revolution, the biotech revolution, the Internet revolution, the revolutionary idea that young people can start companies, the social media revolution, and the blockchain revolution. All of these phenomena have gone global. During the same time span, what phenomenon that germinated in New York City has gone global? Ok, I'm not being fair. The financial crisis did start in New York.

To be even fairer, many of the great revolutions birthed in San Francisco were started by people who were shunned by the semi-tolerant Bay Area culture. Burning Man is an example. If it hadn't been for a city cop telling Larry Harvey to take his Baker Beach fire hazard on the road, he never would have discovered Black Rock City.

Black Rock City is not the kind of place you would have discovered on your own.[146] Not surprisingly, the federal government—an institution that comes to mind for some people when they think of "The Man"—directed Larry and his group of wayward libertarians to a location that would be difficult to find and was devoid of livability to create his experimental civilization.[147]

146. If you type "Black Rock City" into Google Maps, the search engine thinks it is near Blowfish Sushi. A blowfish, incidentally, is an aquatic beast known not only for being poisonous but for its artistic talents. Despite some of David Best's better attempts at building temples using complex geometric designs at Burning Man, David Attenborough has called the blowfish "the greatest artist of the animal kingdom," due to the males' unique habit of wooing females by creating nests in sand composed of complex geometric designs, according to Wikipedia. Also, in case you were wondering, Blowfish Sushi, which relies on precision knifing to protect patrons, so far has maimed no more diners than Benihana.
147. The federal government hosting a burning of an effigy of itself would be akin to King George III hosting a wild party for the signers of the Declaration of Independence.

To get to Black Rock City—a fictional address in the mold of John Steinbeck's Eden—you first head east on Highway 80 from Reno. It quickly becomes clear, extrapolating from early numbers, that you will see fewer than ten living creatures over the next seven hundred miles. The utter barrenness has a way of bringing out the worst of one's neuroses. If I blow a tire and careen into a ditch, how many years until someone finds my remains? Are those vultures I see ahead or floaters in my eye? How many unreported in-laws did the Donner party have to dismember for food while crossing this vast desert *before* they got stuck at the pass?

Fortunately, before the mind can wander too far down this road of grizzly desperation, you are instructed by the Burning Man manual to make a 90-degree left turn off the freeway. That's correct: drive perpendicularly away from your last connection to modern infrastructure toward a minimalist horizon. Within minutes you are without cell reception, GPS signal, or your earlier courage about this whole undertaking.

The hours pass. In the totally blank desert, you are searching for a town called Empire where you can fuel up, and when you get there you realize it's an empire about the size of a gas station and everyone there is wearing no clothes. At this point, the Burning Man manual says you have farther to go before you get to the gates of Black Rock City. Parched, increasingly delirious, and self-conscious about having overdressed for the occasion, for the first time I wonder if I should turn around.

Had I turned around then, or later at what I thought were the Gates of Hell, I wouldn't have been able to close this story with this final thought.

Human endeavor creates value. It is what builds beautiful things and what solves problems. It is the basis of our faith that humans will eventually, even if at some distant time, solve everything—whatever that means. I'm going to go out on a limb and say that the time horizon for this happening is ten thousand years, which is a way of saying—like that drive into the desert—we have no idea how long this voyage will take.

However, a more interesting question than "when" is "what." *What* will we do after we solve everything—including longevity, space travel, world peace, and interdependent stakeholding? We will probably then do exactly what we were doing until approximately ten thousand years ago: gather around the fire, tell stories, and dance. This is what humans have done since the beginning of time and will do until the end of time. It's the enduring human story that stretches in both directions outside the current 20,000-year Grand Interlude—the only epoch in human history where we got caught up in the belly of the beast. Will life be dull thereafter? Hardly. There will always be the love story, and that story never ends.

Of course, we are about as aware of living in the Grand Interlude as fish are aware of living in water. The Grand Interlude is all we've ever known, so we think this is normal. Burning Man, then, is that rare glimpse into the timeless, inaccessible life beyond the Grand Interlude. If you haven't gone yet, I'm not going to tell you stories about it or tell you to trust me. But I will tell you this. Being in the otherworldly place called Burning Man *will* feel like being home. Sublimely. Now close your eyes and imagine what it means to go home to a place you've never been.

Welcome home.

The Closer We Get, the Farther Away We Are

There is a scene in Monty Python's *The Holy Grail* where Sir Lancelot charges toward a castle while unfazed guards size him up while he's far away on the horizon. The scene is amusingly cut in a way that makes Sir Lancelot, in hot pursuit of the heroic deed of finding a mythical chalice, appear to be getting farther away the closer he gets. This *mise-en-scène* also captures the essential paradox of progress: the closer we get, the farther away we are.

Over the course of American history, social, economic, and political progress against power structures has been enormous: the Nineteenth Amendment, the Civil Rights Acts, etc. An unthinkable fantasy not long ago, this century has already produced an African American president and female presidential candidates. The mythical chalice of progressivism would seem to be within grasp.

Yet, like Sir Lancelot's charge, the closer we get to our target, the more it appears to be receding. If anything, while the increasing integration of disadvantaged groups into the American Dream constitutes progress, the resultant dulling of revolutionary ardor has also furthered the domestication of these populations. As we get asymptotically closer to an idealized society without actually arriving, the more subtle the remaining oppression becomes, which makes it harder to even recognize, let alone overcome.

It is this duality of accommodative hegemony that makes it particularly powerful.[148] For example, when the U.S. Army halted the Dakota pipeline, it also halted the pipeline of activists flowing through Standing Rock. This concession stopped the rebellion well short of actually curing the core issue—the exclusive stakeholding by elite institutions that triggered the protests in the first place. Indeed, staging the battlelines along marginal issues, rather than fundamental issues, ensures that this concession to the oppressed is irrelevant in the larger war.

148. Accommodative hegemony is the continuance of tyranny through partial concessions that undermine revolt.

Furthermore, like Nurse Ratched's notebook in *One Flew Over the Cuckoo's Nest,* which misdirects the inmates' enmity toward each other rather than toward the ward's corruption, today's invisible class of elite superorganisms pits its subjects internally against each other to keep them preoccupied. For example, rather than combining their forces to battle the fundamental self-dealing of all politicians, citizens fight among themselves on behalf of politicians for pyrrhic victories on secondary issues.

This type of accommodative hegemony by self-serving power structures has been going on for a very long time. Bobbi Harro calls it The Cycle of Socialization.[149] Roger Daltrey sings about it.[150] It is probably even how eukaryotes got their start, which enabled humans to emerge eventually from the primordial soup.[151]

Nonetheless, we argue that the dehumanizing domestication of individuals by an invisible class of institutional superorganisms constitutes a bug—an evolutionary lag error of human eusociality. The kin skin in the game that served as the genetic algorithm of eusociality within kin hives has been rendered maladaptive in the modern age of high social entropy and genetic melting pots.

Now, our default wiring for eusociality makes us particularly vulnerable to capture, reprogramming, and redeployment in service of self-expanding hives comprised of misaligned individuals. For example, rather than attacking the growth of consumerism that is at the heart of environmental decay, environmentalists today dutifully recycle building materials for the next round of consumerism.

Even the leader in these institutions—the queen bee of modern human hives—is no less an insight-free drone than a drone bee, merely a larger gear in the same machine that everyone is a part of. Natural inclinations and free markets are all it takes for everyone to play their microscopic part in The Man, The Machine, or The Beast.

149. See https://depts.washington.edu/geograph/diversity/HarroCofS.pdf.
150. See https://en.wikipedia.org/wiki/Won%27t_Get_Fooled_Again.
151. See https://en.wikipedia.org/wiki/Symbiogenesis.

From the perspective of The Beast metaphor, humanity was long ago swallowed whole and became cells in the body of The Beast. Unlike cells in our own bodies, where the interests of the cells are aligned among themselves and with that of "the system," in the case of The Beast the interests of individual humans are not aligned with each other or their larger host.

As a result, human civilization today is little more than the sum of all principal-agent problems that emerged everywhere in the post-kin-hive era of human sociality. The dislocation of inclusive fitness was the first-order event. The ubiquitous spread of the principal-agent problems was a second-order event. The tyranny of institutions over the people was a third-order event. Declaring independence from tyranny, as America's founders did, was a fourth-order event. Ayn Rand writing *The Fountainhead* and declaring that institutions are inherently flawed was a fifth-order event.

In all systems, addressing high-order symptoms instead of the first-order causes promotes even higher-order issues. A symptomatic remediation not only distracts from the root cause of a problem but serves as another nail that accommodative hegemony hammers into the coffin of dehumanization. We could make the case immediately for ceasing all "nth-order" reform movements, besides the first-order reform movement of inclusive stakeholding.

Or, we could simply keep charging headlong, as we currently are doing, toward the ever-receding Holy Grail of progressivism.

Part Two

CHAPTER SIX

The Looking Glass Moment

Reader's Guide to Chapter Six

When commodification becomes pervasive across all spheres of life, we lose reference points to orient ourselves. "Nested Dystopia," the opening essay of Chapter Six, orients us to this reality.

The ensuing cluster of six essays addresses our consumer culture: "Fauxism," "Casino," "The Numerator," "Chairlift," "A Future with Fewer Features," and "What's Up with Punk? (Why Creativity Dies)."

The next group of four essays explores the wider commodification of our institutions: "Guns, Roses, War, and Peace," "Second-Oldest Profession," "Alumni Office," and "Domestication of the Environmental Movement." The subsequent essays—"Baseball or Moneyball?" "Other Status-Games Industries," "On Trade," "Heads I Win, Tails You Lose," "Real Estate," "Remission," and "Money Illusion"—relay other games that are played on us.

Their collective hegemony is so complete—as discussed in the final essay in Chapter Six, "Out Through the 'In' Door"—that, even when we do try to get out from under it, it's not possible to tell if we are actually awakening from a dream or dreaming that we did.

Nested Dystopia

When Alice steps through the Looking Glass, she enters an inverted world. Like a reflection, everything is reversed, including logic: walking away brings her closer, inanimate chess pieces are alive, and fictional characters actually exist.

As we leave our eusocial hives for dissocial ones, we, too, step into an inverted world. Without vested interest in others, self-interest prevails, and people subvert, rather than serve, each other. Superorganisms subvert, rather than serve, individuals. A mirror is just a mirror—another tool for vanity.

Indeed, the closest thing we experience to Alice's portal is not the mirror on the wall but the front door to the outside world. Walking out that door is when we exit the last vestige of eusociality—the nuclear family—into a world of dissociality—a rat race being run on a massive scale.

But, unlike Alice, we are unlikely to notice much as we cross the threshold. In contrast to her Wonderland, fictional characters in our world don't talk to us, words aren't read backwards, and chess pieces aren't alive.

Yet that's only true because our Looking Glass turned out to be *two*-way. On the one hand, being in your home may feel like you are living the American Dream, yet things also don't feel quite right. That's partly because dissociality has stepped into our eusocial nests and begun to take over every inch of our lives, including inside our homes. And even when dissociality is not *in* the nest, it's in the nest—listening and filming, maybe even adjusting our thermostats.

To call our homes nests may feel like a bit of a stretch. In the old world, we would have built our home with lumber we cut ourself. But do today's homes sometimes seem like cookie-cutter cages built to line the developers' pockets? After all, don't we provide the bacon they need to bring home?

How we live inside our physical environment also doesn't feel quite right; it is as if distributed social algorithms are running people's lives. Kids bring work home from an industrial education complex that prepares them to be line workers in a larger system of industrial complexes, from which they will one day bring work home.

When the ever-escalating homework finally gets done, the food industrial complex feeds its denizens, hijacking parental love to make it appear that the food will serve their children well, instead of the shareholders. After dinner, the system lures the denizens like lab mice to in-house two-way looking glasses where they can suckle from the machine, one Pavlovian mouse click at a time.

If you take a step back, you can see the inversion. The nest is now a server in a decentralized, interconnected network of servers. Each server consumes energy in a way that ultimately allows individuals to mine the dollar alternative to Bitcoin through proof of work and stake—a process that keeps the gears of the larger system running while the individuals in the system run in place.[152] Or, we can keep pretending that this is not what is actually happening.

Our purpose of going down this rabbit hole is to explain the fallacy of the analogy of your front door being a portal between two different worlds. Whichever side of that door you are on, you are still seamlessly in lockstep with the same global march of dissociality. Even your own home—your nest—is just a nested dystopia.

Only by stepping out of the realm of your entire known experience will you notice that everything is running backwards. Counting down the calendar years of a dissocial civilization has been sold to the world as forward progress. That viewpoint comes to life when noticing the directional reversal of relationships from service to subversion in every dimension of the human experience, resulting in dehumanization.

Once you line up your personal frame with this outer frame of reference, you can start to make better sense of the senseless things going on everywhere in the illogical landscape that is our reality show—where fake characters do talk to us in gibberish, a rip-off stand is unironically called a concession stand, and we all serve as pawns in a larger game.

Part Two opens the door into this bizarro world.

152. It is as if the system is using humans the way humans use machines to create Bitcoin.

Fauxism™

Editor's Note: As motherships replace mother, wisdom recedes.

No almanac of cultural commodification would be complete without mentioning its mothership—Disney—where everything possible is leveraged performatively for consumption. Who else would commercialize the sanctity of a happy childhood by trademarking the phrase "The Happiest Place on Earth™"?

In some ways, however, the transparency of Disney's commodification also makes it "The Safest Place on Earth." Unlike other markets—politics, corporate management, Hollywood, etc.—that select for the actors who are the best at pretending they are not acting, Disney does not pretend to be what it's not. Its pretension is bona fide.

You already know all this, of course. You were probably dragged to Disney World as a child out of parental guilt and then dragged there again by your child out of parental guilt. You accept the cycle of life.

But just in case you're the type who can't find the glasses you are wearing on your forehead, Disney makes its vision cartoonishly obvious. At a time when storytellers were rushing to make fantasy appear real, Walt and Roy Disney went the other way; in 1928 they created Mickey Mouse. Perhaps after the serial disenchantment of reality embodied—and disembodied—in Fauvism, Cubism, and Surrealism, this was the next step in high art.

Play along with me for a moment and put Edvard Munch's *The Scream* next to the original sketch of Mickey Mouse.[153] [154] Seen in this juxtaposition, it becomes apparent, which is the higher artform in disguise. By the time Mickey Mouse hit the scene in the late 1920s, the existential angst of the art community had cornered itself into an impossible contortion (see the first image). By contrast, the

153. See https://en.wikipedia.org/wiki/Mickey_Mouse
154. See https://en.wikipedia.org/wiki/The_Scream

second image—the same angst now masked as a happy-faced rodent captaining its own destiny—is the release valve for the accumulated pressure to conform to nonconformity.

That is to say, rather than fighting the overlords of dystopian reality, the cornered mouse of art waltzed past its unsuspecting guardians by dressing as a Mickeyfied version of itself. There was, frankly, no other way for real high art to emerge in the days when elite gatekeepers were policing the field; subversion was the only way to go to market.

In fact, going to market *was* the art form. The desire to focus on the money had to be genuine. That is why, according to Walt Disney, Mickey was drawn one finger shy of a full set in each hand. He estimated that not tiring the hands of his laborers—who had to mass produce many frames per second worth of drawings—saved his studio millions per film.[155] Why cut corners when you can cut out an entire finger? Similarly, that Mickey's features are drawn as circles seen from any direction wasn't some poppy-induced artistic statement. It was a financial statement.

But before this financially modeled Mickey could go on to star in 130 hand-drawn films, he would first have to prove himself in front of a live fake audience.[156] Although it might seem hard to believe now, at the time, adding sound to make-believe moving images was an unproven commercial idea. So Walt Disney had his employees perform live music behind the curtain where the movie reel was projected to create a real feel for what recorded music might sound like.

Resisting the temptation to name this emerging novel art form Fauxism™—not to be confused with its predecessor, Fauvism—was part of the success of this new direction for commercial art.[157] Fauxism's authenticity lay in how seamlessly it mirrored what was happening in culture. In the real world, a "show" had become not only what you revealed but what you performed. An "act" had become not

155. See https://en.wikipedia.org/wiki/Mickey_Mouse.
156. See https://en.wikipedia.org/wiki/Steamboat_Willie.
157. Fauxism is an art movement wherein pieces of artwork are created entirely for the purposes of commodification.

only what you did but what you pretended to do. An "image" was no longer just a photo but also a brand. You get the picture.

It turned out that Disney's first live test of fake live music performed behind the curtain wowed the crowd, and soundtracked cartoons soon took off. Then, two years after he was born, Mickey spoke his first words: "Hot dogs! Hot dogs!" he said while selling hot dogs at a carnival. If this sounds like the potential business model fractal for Disney—peddling fake meat at a dystopian clown show—let that sinking feeling sink in for a moment.

The Disney fractal is also a fractal for overall society. Everything has become a commodified version of its original purpose. Doors are opened for tips. Kids study to get into college. Events are entered for their Instagrammable value.

If that event happens to be the Super Bowl, winners of the fantasy match between our proxy uncles will most definitely sip preshipped champagne, don prefabbed hats, and spontaneously look into the camera and victoriously yell what we are all programmed to expect: "We're going to Disney World!" As if they couldn't have just paid to get in, just like the rest of us.

But let's face it, we all know that no one plays the game of commodification better than Disney. For sociology graduate students to write new theses about Disney's commodification would be tautologically redundant: "Disneyfy" is already a defined verb that means "to create or alter in a simplified, sentimentalized, or contrived form or manner."[158] The word "Disneyfication" has already achieved the cultural status of having its own Wikipedia page, where it is defined as "the commercial transformation of a society to resemble the Walt Disney Parks and Resorts."[159]

The social significance of Disney, then, is not that it represents a grand theft of authenticity; it's that it represents the true yardstick by which to measure all other forms of commodification. Like Burning Man, the Land of Oz, or Alice's Wonderland, it helps us see our world more clearly by revealing what it's not.

158. See https://www.dictionary.com/browse/disneyfy.
159. See https://en.wikipedia.org/wiki/Disneyfication.

Without it, it might actually be hard to know that we are getting duped across the board. Having a gold standard helps—even if it's just a paint job.

For example, for the castle-like four-star hotels to live up to Disney's impossibly high standards of commodification, their employees would have to greet you wearing animal costumes. For the department of motor vehicles to do so, its employees would have to wear a smile. These are just Mickey Mouse operations compared to the real thing.

In the grand scheme, all painted brick roads can be followed back to our original sin: not updating inclusive fitness with inclusive stakeholding in order to maintain our eusociality. When we shift from being served by those who love us most—such as our moms—to being served by those who love themselves most, the Disneyfication of our experience is the ultimate price we pay.

As such, there's no act in Disney's act. Their plans are all disclosed in SEC filings. The workers' malcontent behind the costumes airs regularly as dirty laundry.[160] We know their movies will open with child abandonment.[161] Indeed, this latter theme that we were all exposed to as kids—like Mickey Mouse's first message—is perhaps the company's most devastating message of all: remove mother from your illusory happiness and come join us on the mothership.

160. See https://nypost.com/2019/07/15/disney-heiress-says-employees-unhappy-with-working-conditions/.
161. See https://en.wikipedia.org/wiki/Bambi.

Casino

Editor's Note: Once a hive shifts from eusociality to dissociality, all bets serve the house and not the players.

Perhaps you've heard someone say, after a Las Vegas trip, "I lost money, but I had fun." Think about that sentence again. Do those seven words make sense? Microeconomics say people earn money by performing tasks for The Man that are so not fun that you couldn't pay anyone else to do it at that price. Why, then, would someone give it all back to The Man over five cocktails and call it "fun"?

It makes you wonder if the meme "I lost money, but I had fun" was a winning entry in a marketing-slogan contest run one drunken night at the Pair-o-Dice Club on highway 91—where Vegas got its start.

The slogan works. It is parsimonious, devoid of distractive adjectives, and easy to remember, unlike the rules from the book on how to count cards during a game of blackjack after five drinks. But when you can't fully explain to your spouse why your pockets are empty and little Johnny now has to pay for college himself, at least you have the go-to alibi, "I lost money, but I had fun."

What happened, of course, is that your money stayed in Vegas—the unedited version of another popular marketing slogan.

But, in your mind, it is hard to reconcile exactly how that happened. That's because you were pretty sure you were up the whole night and kept telling the well-made-up Avon lady next to you all about it. Or maybe it was the one wearing sunglasses indoors, who made up that story about once touring with the Chainsmokers as a backup singer. Wait, never mind. That was the one you told to stop chain smoking in your face.

Through all the haze, the reason you were sure you were up all night is because you remember multiple occasions when you were in front of a machine that kept paying out large sums of money. You did it exactly like you always do: insert member card, push button, and machine pays out money. The only thing you can't explain is why no bells went off like you see when other people win jackpots. You also can't explain why you have a text message from Wells Fargo saying you have overdrawn your account.

Fear and self-loathing about possible bankruptcy, however, won't stop most people from going back to Vegas. For millions it is America's Mecca—an oasis of transparent capitalism in a land of mirages. And don't let the region's barren desert fool you into thinking the place doesn't have a rich history. It's rich in uranium. Even in this land of make-believe, it's hard to believe that 928 nuclear bombs have been detonated—sometimes in front of huge viewing audiences—at a site just a one-hour drive from Las Vegas.

In nearby Hollywood, pilots that bomb with test audiences have a short half-life. But this one didn't. The atomic program ran from 1951 to 1992, making it the second-longest-running program in the region's history, after *Les Folies Bergère*.

Meanwhile, in authentic Vegas style, the city packaged those atomic spectacles as tourist attractions. The Chamber of Commerce distributed calendars promoting detonation times and premier viewing spots. Binion's Horseshoe and the Desert Inn lured guests with promises of north-facing vistas and "Dawn Bomb Parties." Sands ran "Miss Atomic Energy" pageants in which women were (barely) dressed as mushroom clouds.

The question is, why do people willingly pay money to do things that are all but obviously not in their best interest? In the aftermath of Hiroshima and Nagasaki, there were no illusions about the horrors associated with radiation. At a time when propaganda about the risks of the bomb was supposed to scare people into building bomb shelters, Vegas was able to use the public's strange love of real-life detonations to sell weekend vacation packages.

That may have set the all-time rock-bottom low of commodification through capitalism. Once the town passed the test of selling a nuclear blast as a family holiday, it was clear to Vegas that all bets were off. They could now sell practically anything: drugs, women, dreams—you name it. Steve Wynn did.

But, what is less clear is why people buy into places that so overtly sell them out. Why do they risk their lives and livelihoods and spend their fortunes to help others make theirs?

This brings up the larger point. The casino is perhaps the world's most honest example of accommodative hegemony. A house sets a few rules that ever so slightly favor the house. After that, all it takes are natural inclinations and free markets for individuals to voluntarily transfer economic power from their own bank to the house's. Everyone knows the game being played, but people play the games anyway.

For the house, the game is real. When it comes to service, attention to detail is part of the subversion. It's not just the dopamine-feeding visuals, sounds, and tastes that everyone knows about. It's also the lesser known tricks of the trade such as the green color of gaming tables and paths to bathrooms lined with sirens that can wreck your financial future. The path out of the building is even harder to find; like shopping at IKEA, it's hard to leave when you can't find the door.

Meanwhile, anything that might help you avoid personal bankruptcy is blatantly absent. While alcohol and other debilitators of prudence flow freely, the coffeemakers and caffeinated beverages that stimulate rationality are nowhere to be found in casino hotel rooms.

Seemingly nothing has been overlooked when it comes to finding human behaviors to exploit. But, just in case there is more to learn about the human animal, one-way cameras overlooking the Pavlovian proceedings are spaced every few feet apart.

Perhaps we can widen our own lenses and reflect on the broader proceedings. Like casinos, all of our social, political, and economic institutions—politics, education, healthcare, media, Wall Street, etc.—are also houses of accomodative hegemony to one extent or another. Like casinos, they write rules that favor the house, then let free markets and natural inclinations take over. Like casinos, power flows from individuals to the house under the illusion that you have control.

Unlike casinos, however, you may not realize that you are playing their games.

Whereas casinos are overt, others subvert. The lack of transparency is a clue that these institutions have yet to evolve fully into voluntary forms of tyranny.

More advanced forms of accomodative hegemony, like casinos, come with accommodations. It means people want to spend the night. Less advanced forms come with clocks. It means people might be trying to get out: 9-to-5 workers, schoolchildren, and inmates. That casinos don't have clocks is an indicator of voluntary tyranny. The bells and whistles keep people in their seats, rather than making them run for their lives.

And that brings us to the final point. Inside the casino, the game is already over. Outside, however, it's not over yet. While the decks may be stacked against the people, there is still some hope. That people are willing to run away from tyranny, unlike from accommodative forms of hegemony like casinos, is the sign that we might still stand a chance.

The Numerator

I needed the numerator. So I asked the sunglassed young man wrapping a towel around my wife's ankles how many people had died while bungee jumping here. I wasn't asking because I had an insurance policy out on her, and we didn't have kids to orphan yet. I just wanted to know.

"None while we've been sober," he answered, while touching up a knot that would hopefully not slip with nine Gs of tension.[162] It was a reasonable answer. He probably didn't know the number, and I definitely didn't want to know it. But I was glad he didn't look up from his task while answering me. I didn't want to see if his pupils were dilated.

If I'd really wanted to know how many of his pupils had died of late, I should have asked it three months prior, when I was booking bungee jump for my wife's birthday gift. To ask it now, while with her on the edge of the Kawarau Gorge Suspension Bridge looking out over the Kawarau River, standing next to a ski bum working his backup job, was just asking for it. I got the answer I deserved.

Polite nervous laughter followed. That was a dead giveaway of how I really felt about failing to stand up to this bully, who was now putting my wife on a hook. Meanwhile, succumbing to his deft sarcasm, I had let him off the hook. That he uttered his reply so dryly suggested that my needy question was, to him, just a line item on the FAQ.

It worked. His sarcastic retort to my snarky question gave him just enough time to execute the transaction before I could recover from his deke, cancel the proceedings, and ask for a refund. Over the gorge she went.

In the current post-modern landscape, where humans are the apex alphas, the biggest risk we face every day is counterparty risk: that is, each other. People put food on their table by offering something for trade. Given the nature of counterparties, however, there is some degree of imbalance in every one of these

162. See https://www.rigidlifelines.com/blog/entry/bouncing-bungee-bodies#.

transactions. In the case of adventure tourism, you offer your life, and they offer you dopamine and witty repartees.

This is all part of the carnival act. When you are the mark, you are the last to know it. For example, if you were paid as a carnival actor to take leaps of faith on a trapeze, there would be a safety net. But when you are the one paying for the privilege of taking leaps of faith, your signed waiver is *their* safety net.

Why we keep falling for this is one of life's mysteries. But that question doesn't concern tour operators. If it will put food on their table, they will put your children up for food. That's how, in 2015, I got suckered into letting the kids get surfing lessons during tiger shark mating season in Hawaii.

I had learned from personal research that, between July 31, 2013, and April 29, 2015, one person on average suffered a shark attack every six weeks in the Makena area, where our group of twelve was unfortunately vacationing.[163] Something told me I had folded better hands in poker, even when losing the hand wouldn't have required emergency surgery.

To better appreciate the odds, I looked out at the beach and estimated that, as far as I could see, about one hundred people per day were going into the water. Since we were staying for a week, if we went in the water every day, the probability that any one of us would be bitten by a shark that week was a cheery 1 in 600. The probability that *someone* in our group would have this experience was 1 in 50. After the eons of evolutionary work our ancestors did so we would become apex alphas, was this how we were supposed to roll?

A more refined analysis of the data was not encouraging. The majority of attacks occured in shallow water, in broad daylight, and in popular beach areas. And this wasn't just a theoretical risk. Every time I refreshed the website that tracks the tiger sharks tagged by the University of Hawaii, I could see four to six females and one male bull reading the menu in Makena. When I began calling around local surf shops to inquire, one person who picked up the phone confided, "I surf, and none of the local surfers I know in Makena are allowing their children in the water." Click.

163. See https://dlnr.hawaii.gov/sharks/shark-incidents/incidents-list/.

You wouldn't have learned any of this by talking to the resort's activity guides. Every time we started a group lesson of any activity that involved legs being in the water, I shared my findings about recent shark attacks. Every time, the guides gave the same rotation of responses: "Only happens when you are alone," "Only happens in deep waters," "Only happens early in the morning or later at night" (i.e., non-business hours), and, my favorite, "I've never seen one happen." Their real occupational hazard, apparently, was dealing with morons who ask for the numerator.

Such is the reality of adventure tourism distorted through the prism of incentives and conflicts of interest. Whether it's risking swimming with actual apex alphas, getting impaled while skiing, or hitting the Kawarau River face first at terminal velocity, Americans love their dopamine fix. And whenever money and lives are both on the line, there is a sobering possibility that an entrepreneur will be there to serve as a counterparty for your carnival act.

Buyer beware.

Chairlift

Editor's Note: "Chairlift" is an ode to the particularly strange sport of racing down the mountain on a slippery slope.

If I were to tell you about parents who put their children on a park bench, unharnessed, six stories in the air, you wouldn't hesitate to call child-protection services.

But you put that same aerial bench high in the mountains where there is less oxygen to perfuse the brain, it's called a chairlift.

Here in Winter Wonderland, 8,217 feet above sea level, there's no helicopter parent in sight, and it's actually the children who are hovering acrobatically, trying to keep their center of gravity on the bench despite the fifteen pounds of unwieldy cantilevers boot-strapped to their feet. Their ski outfits were optimized for qualities such as hue and number of pockets, but not for gripping a slippery surface that sways during wind gusts.

You sent them up alone to help nurture their independence, and your children are now among an unvetted grab-bag of folks about as diverse as the colors of snow. Your children might be sitting next to other unrestrained little children with no life experience or with twentysomethings on furlough from pot farms, who at least are paying more attention than the corporate raider on a conference call barking insider information, which your children are now obliged to report to the SEC. You have done your due diligence in every other area of your life to make sure your children are not with the wrong people, but no background checks are done when everyone is wearing a ski mask.

Thankfully, none of them is driving the aerial chair because this one is self-driving. Actually, to be less inaccurate, the chair is being driven by another twentysomething—the out-of-sight assembly line manager whose job is to make the entire conveyor belt of volunteer trapeze artists come to a sudden and swinging stop if a paying customer not yet dangling in the air mistimes his or her moving merger with this purportedly non-lethal electric chair.

You are comfortable with all of this because you yourself have been on this high-wire bench many times without incident, as you are doing right now on a different part of the mountain. After all, you would never let your children do something you wouldn't do, such as count the number of orange body bags being dragged hurriedly off the mountain. You trust that the person on the sled is not moving because they are strapped in and cocooned in the safest position they will be in all day—supine and in contact with the earth's surface—and not because they are dead.

As you rubberneck at the body bag racing in a downhill direction, your mind briefly contemplates your own potential upcoming downhill morbidity. Such grisly visualization is, of course, for pessimists, so instead you do the math, and the math is heavily in your favor: most people you've seen going in the counter direction were not travelling in orange body bags. It's the kind of deep learning, shaped through evolutionary eons, that allows a herd of proud gazelles to semi-confidently waltz past a pride of lions on the Serengeti. It's also the same math performed before a person decides that it is statistically safe to dress like a seal and bob on an appetizer wafer thirty-five miles from the Farallon Islands.[164]

You kind of like your personal odds, so you return to thinking about the poor schmuck whose day—or possibly days—of walking without a cane is over for a while. You think of some silver linings, such as that the messed-up joint at least got iced immediately while buried in the snow. But it's a long ride, and your mind drifts again like a wayward flake. Now, there's a part of you that wishes your friends could see you take a bad fall in which you hurt nothing physically, but, out of an abundance of caution for your third-grader ego, your friends repeat all the right buzzwords listed in the resort's liability handbook to earn you a dashing snowmobile ride to the sanctuary of a fireside hot chocolate.

But this is the real world and you have no such friends, so you start drifting deeper into a vague memory of Googling research papers, funded by the National

164. See https://www.sfgate.com/news/bayarea/article/17-Foot-Long-Great-White-Shark-Bit-Diver-Near-13336137.php.

Ski Area Association, showing that downhill winter sports are no more dangerous than ping pong. And, before you can come up with the five worst injuries ever suffered by a ping pong player, the aerial bench lurches, your center of gravity tilts a bit farther forward, and you are plopped gently at the top of the mountain—thus ending the safest part of your ski vacation.

A Future with Fewer Features

In our house, the process of turning on the TV is a little like launching a nuke. Two or three people are involved, and no one person knows the full sequence of buttons to push. If guests stay over and futz with the controls, forget it—we have to reset the system from scratch.

When did everything get so complicated?

I've traveled for business most of my adult life, so you'd think I'd feel like a pro by now. Instead, I feel like a moron. In some hotels, just trying to turn on the lights has become trying. What's with these light-control panels with dials and knobs? No, I didn't want to move the shades up and down. Why would I want to hold down a button to see if the lights will get less dim? What happened to the old-fashioned on-off flip switches? I know they are trying to save electricity, and it's working. More than once I gave up, unpacked in the dark, and went to bed.

They are also trying to save water. Turning on today's fancy showers now requires passing the prerequisite exam of turning on the lights. Gone is the pedestrian two-knob system—you know, the ones you turn to mix the flow of hot and cold water—that you could operate without needing to put on your glasses. You now have to wear glasses to take a shower, and even then it's hard to decipher the dashboard of knobs, handles, and dials. In the worst cases, you can't reach the dashboard while standing safely outside the shower. So, the sequence goes like this: get naked except for glasses, stand under shower, and pull handles like you are in Vegas. It usually works out fine. Other times, less so—like the time an arctic blast waterboarded my groin from a second showerhead.

Everywhere I go now, people performing routine tasks look like they are taking a Rorschach test. Once upon a time, you parked your car, put change in a meter, and walked away whistling. Now the cars park themselves, but paying for it takes nine steps. First, you scan the horizon for a kiosk, which is rarely only nine steps away. Second, you speed-walk to the kiosk and settle into a queue. There, you patiently wait as the person at the front calls tech support. It's only when you get to the front that you realize everyone ahead of you has left the kiosk to retrieve their license plate number, and they are now standing back in line behind you.

In 2004, Barry Schwartz wrote the book, *The Paradox of Choice—Why More Is Less,* in which he explained how giving consumers too many options was causing them stress and paralysis by analysis. A 2010 meta-analysis of fifty studies disputed Schwartz's claims, while a 2015 meta-analysis of ninety-nine studies affirmed them. During the eleven years the field was paralyzed by these meta-analyses, the stock of Apple—the company that mastered the art of giving consumers no choice—rose from $2/share to $112/share.

Before Apple, buying a computer involved walking into Best Buy knowing what you wanted, then leaving empty-handed and dazed by options you didn't know existed but now craved. The experience at Apple stores is the opposite. Their stores are cavernous, setting you up for that guarded feeling of walking into an indecision nightmare. But, as you peruse one computer after another, you soon realize that every computer on the table is identical, like your image in a hall of mirrors. You quietly cheer because you know you can't lose when you can't choose. What Apple is really selling is reassurance—that elusive feeling that you are not a failure and not a bad person. Who knew that giving consumers no choice—a key feature of communism—would be the new capitalism? At least you no longer have to choose between these two economic philosophies.

Or, so it seemed. In the real world outside of Apple stores, people are being crushed more than ever by "option obesity" and losing control of everyday mundane tasks. No one has time anymore, and everyone is running out of it. For those of you on consumer research teams, here's a little secret: simply put, the next great feature revolution is the lack of them. People long for a simpler time, so give it to them.[165]

Oh sure, for people who absolutely want to maintain control over all permutations of options—and apparently have time to read user manuals at a time

165. Meanwhile, oversimplification, as a concept, can also be thought of as a social technology. In a high alignment world, people are biomotivated to use oversimplification to serve; in a low alignment world, people are biomotivated to use oversimplification to subvert.

when no one even reads user agreements—keep the complicated dashboard but color it green. For the rest of us, please include a conspicuously placed large red panic button on all of our appliances that will make devices perform their most commonly intended task, such as turning on a microwave for one minute. We're not always going to have a ten-year-old around to navigate modern technology for us.

Or, stay the course and continue to make all of our lives ever more complicated through technology.

It's your choice.

What's Up with Punk? (Why Creativity Dies)

"If you looked different, people tried to intimidate you all the time. It was the same kind of crap you had to put up with as a hippie, when people started growing long hair. Only now it was the guys with the long hair yelling at you. You think they would have learned something."

—Bobby Startup166

At its outset, punk did to rock and roll what rock had done to hippy culture—saved it from consuming itself. That is to say, music was once more pushed to the edge of chaos: raw, challenging conventions, and then some. In doing so, the scene was cool again.

At the time of reliable lies that were the late 1960s, punk was truth because it didn't beat the generation around the bush. To play punk was to make sound, not produce it. There was no propaganda to decipher because everything was front and center. Initially, none of the artists were commercially successful, which kept reputations intact. "Here is our music, so shove it," is roughly how it went. Onlookers questioned whether it was even music. And that was the point. Uprisings like punk are not supposed to answer to authority.

The genre didn't really start at a particular time or place. Sure, maybe Lou Reed had an early role in it in New York City, but that underground subculture bore no likeness to the buttoned-down Jonathan Richman, and it clashed with the rev-heavy Ramones. After all, discoordination was a feature, not a bug, and to have organized would have defeated the whole purpose.

However, to understand genuinely the mercurial origins of punk—and its eventual fall—you have to go further back, not just to the time of Elvis, James Dean, Jack Kerouac and the way they lashed out at authority but much further. You have to go back to the beginnings of human social evolution.

166. *Philadelphia Weekly,* October 10, 2001; see https://www.etymonline.com/word/punk.

After humans left their kin tribes and formed melting pots, altruistic herds turned into selfish herds. Individuals in dissocial groups, as opposed to eusocial groupings, are bioincentivized to hoard and wield—not provide—power.

In eusocial herds, the empowered protect the weak by putting their own bodies between the threat and the threatened. By contrast, in dissocial herds, the empowered jockey for the central protected position while leaving the weak to the edge.[167] This is how mobs work.

Similar social dynamics govern memetic herds. Whereas different physical individuals cannot occupy the same position, different memes can. When a threat is perceived, the accelerating momentum to the center of the memetic herd leads all individuals to collapse into a single consensus. From a social dynamics perspective, if the eusocial herd is the sun, then the dissocial herd is a black hole.

On that cheery note, let's discuss teen social life.

The onset of puberty was when pre-modern humans left the bosom of their nuclear families to begin their dispersal, find mates, and renew the cycle of life. Since today's culture is more complex, parents sensibly try to keep their children under their roof to house-arrest their development. The tension produced by this evolutionary dislocation is the sociobiological reason teens act like they know it all.

But that's just the beginning of teen angst.

After emotionally cutting the cord to their last attachment to eusociality—their unquestioning trust of their own parents' loyalty—teens enter the mother of all industrial complexes known as high school. There they will be unprotected by the Fair Labor Standards Act of 1938, intended to ensure that, when young people do work, it does not jeopardize their health, well-being or educational opportunities. But, if the stress of dealing with the system isn't enough, teens have to deal with each other. And that's where it really gets rough.

There is a reason why social dynamics in the real world are often compared

167. In the case of the GameStop saga of early 2021, a Main Street mob challenged the wolves of Wall Street to a street fight. The trading platform Robinhood, which sells users as products to buyers on Wall Street, was portrayed as a wolf in sheep's clothing, seemingly serving the people's side against powerful institutions while allowing the latter to front-run the former.

to high school social life. The latter is an under-the-microscope microcosm of the former, a pressure cooker of peer pressure. The beginning of school is a social Wild West, one large dissocial memetic herd decomposing into many sub-herds: jocks "jockeying for chicks," cliques competing for clicks.

The rules of the game are clear at all times: be cool, or be cast out.

The pressure to conform in high school is the beginning of the suffocation that leads to another archetypal social byproduct: the rebel. The latter character needs no introduction, since rebels are the subject of much fanfare. They are in fact the easiest species of high school student to spot, as they occupy the fringes of the campus, social circles, and academic charts.

What rebels might sometimes lack in scholastic aptitude, however, they make up in attitude. Rebels come in all shapes and sizes—we're talking about hair and accessories here—and a lack of convention is their convention. What they share is a common ethos: rejection of authority and peer pressure.

In doing so, rebels manifest the buds of something that conformity sporulates: creativity. The self-mutilating accoutrements and radical self-expressions of teens—alcohol, cigarettes, drugs, tattoos—reflect a kind of physical graffiti on what may feel like the last bastion of personal sovereignty. Some even take it further, as an autonomous collective of hackers taking on power structures. They know the price of these forms of expression is high, but they act out because they feel the price of surrender to conformity is even higher.

They know this because, to varying extents, they can see it in the lives laid before them. They see adults listlessly meandering in the industrial complex, devoid of their own ideas or the courage to express them. To be in a corpocracy is to be in the belly of the beast, where there are few upsides and many downsides to not participating in the rituals of mediocrity, complacency, and bureaucracy. Without mutually vested interest in each other, everyone from top to bottom feeds from the trough of the corporate treasury, rather than feeding into it. The emphasis is on performative acts, not performance. Self-interest grinds the gears of creativity to a halt until rigor mortis sets in within the corpocracy.

In that setting, creativity has no home. Memetic heterogeneity is a threat to the consensus, something to be shamed and possibly punished in enough of a public way to discourage others from following suit. Eventually, all edges of creativity are dulled and rounded until "Toyota Camrys of everything" roll off the assembly lines of the entire industrial complex—meek, safe, devoid of spirit, mass-produced according to quality-assured moulds like their dehumanized creators.

In other words, the same forces that make teens performatively mimic each other's dance moves on Tik Tok in order to promote themselves also makes American cities look similar to each other and make sporting events across those cities perform similar fan rituals and soundtracks.[168] It is also what keeps scientists and institutional leaders in their mainstream lanes rather than taking bold risks.

The conforming industrial complex of everything subsumes the art and music worlds too.

This is where punk came in. During a time of disillusionment of the late 1960s, punk took the opposition position, not only to the commercial, over-produced, self-aggrandizing culture of rock but to the larger culture itself. It erupted onto the music scene in loud tones and lyrics that were tone deaf to the bourgeois sensibilities of the mainstream. It may have taken years for punk to graduate from its underground roots, but when it did, it took the world by storm.

By then, however, the sentinel signs of downfall were evident. When 1976 was declared year zero for British punk, it was already clear that the whole thing was stillborn. Infighting as to who was authentic and who was a poser—*poseur* for those that want to act out the word while pronouncing it—replaced the fight against the establishment. None of these skirmishes mattered. By the time the word "authentic" enters the vernacular of any field, all is lost anyway.

The establishment, meanwhile, had a field day. Like the Beatnik and hippie cultures that preceded it, the culture of punk became a neatly packageable commodity. From copycat bands on vinyl to vinyl-leather outfits in malls, the

168. See https://www.theatlantic.com/technology/archive/2020/08/why-every-city-feels-same-now/615556/.

marketable lifestyle sold out stadiums, as well as its own soul, before ending up in the markdown racks at K-Mart. Before long, punk had turned into the very thing it thought it was fighting, and it never stood a fighting chance. It would be followed by other fleeting insurgencies—grunge and hip hop, for example—that would temporarily carry the torch of *avant garde* counterculture.

The self-consuming nature of these insurgencies is part of a larger story of all reactionary movements. All are Band-Aids on the symptoms. Neither the conformist nor the rebel has a clue that their two-sides-of-the-same-coin identities are shaped by forces outside of what are spoon-fed to them by schools and revolutionary tracts. The true force behind both conformity and rebellion is the transformation of eusocial human hives into dissocial ones. Invariably, this force makes every rebellion eventually eat itself, the victim of the same contrivance it was trying to reject. The chorus section marching in unison to the line "we don't need no education" in Pink Floyd's "The Wall" demonstrates the recursive irony of rebellions and its inevitable takeover by the black hole of conformity.

It is thus that at Burning Man a man is shamed out of wearing his collared shirt and dress pants—the "I am middle-management company-man look." The unspoken pressure to outfit one's misfit in a certain way—cyberpunk, steampunk, and other hand-me-down cliches of art noir counterculture—reveals counterculture's inability to outrun the very forces it tries to leave behind.

The larger point is this: both the conformist and the rebel are parallaxed offspring archetypes sprung from the loins of the same mother: human dissociality.[169] Unless this latter issue is resolved, the yin and yang of culture and counterculture will forever chase each other.

In the meantime, for young people today who are looking for a radical idea in yet another era when public faith in institutions is crumbling, consider this: you've been conditioned by the system to believe that you have to choose between suffocating conformity or the wasteland of teen rebellion. But there is a third option.

You can be yourself.

169. The birth order is perhaps not unfamiliar: the dutiful first born and the irrepressible second born.

Guns, Roses, War, and Peace

Is the military -industrial complex incentivized to create its own demand? Do purveyors of war self-deal for power? Could self-interest be mitigated with more skin in the game? War psychology is a notoriously difficult terrain—complicated, and sometimes undermined by the fog of good intentions—that Julius Caesar, Leo Tolstoy, Joseph Heller, and many others have explored.[170]

Relevant to our discussion are deeper questions that are even harder to answer. Would war deciders decide differently if they had more kin skin in the game? What if their own children were at risk of being randomly chosen to serve on the front lines? That British royal family members are expected to, but not required to, serve in the military doesn't even begin to scratch the surface of the complexity of competing loyalties to kin and the state.[171]

Among warring eusocial ant colonies, the rationale is simpler; for them, "Long Live the Queen" is more than a gesture of loyalty to royalty—it is their livelihood. To sterile soldier ants with kin skin in the game in their queen, self-sacrifice in battle on her behalf amounts to genetic self-preservation. From her perspective, their loss is hers: imagine the sorrow of being the next of kin to every fallen soldier ant. Even more to the point, the losing queen will be vanquished, and in a welcome-to-the-jungle, sweet-child-of-mine moment, a daughter of the victorious queen will seed the conquered hive with her own progeny.[172]

Heroic deeds by soldier ants are part of a larger story of kin loyalty's influence on war behaviors in nature. Ethnographers have long noted its similar influence on intratribal and intertribal warfare among indigenous human populations.[173]

170. See https://infed.org/mobi/ivan-illich-deschooling-conviviality-and-lifelong-learning/; https://www.vanityfair.com/culture/2011/08/heller-201108; https://philosophiatopics.files.wordpress.com/2018/10/skin-in-the-game-nassim-nicholas-taleb.pdf.

171. See https://www.popsugar.com/celebrity/Can-Royals-Serve-Military-45450435.

172. See https://www.youtube.com/watch?v=qLQLKT8enys; https://www.scientificamerican.com/article/the-hives-of-others-bees-wage-war-across-species/.

173. See https://www.amazon.com/Yanomamo-Fierce-Studies-Cultural-Anthropology/dp/0030623286.

But kin skin in the game can also serve the opposite purpose—as a promoter of peace.

Chief among these rituals is the tradition of royal intermarriages—a practice familiar to those steeped in the lore of European interdynastic nuptials. When former enemies suddenly become 50 percent vested, in the genetic sense, in each child sprung from these arranged unions, the incentive for the two houses to cooperate blossoms. It was the betrothal of Elizabeth of York to Henry the VII, for example, and not guns that truly ended the War of the Roses and brought peace to the reunited royal houses.[174]

Such storied intermarriage is just the tip of the ant hill. The practice of competing factions merging bloodlines to promote diplomacy, expand domains, and form strategic alliances has been found on every continent at least as far back as the Late Bronze Age:[175] Roman emperors arranged marriages between ruling families of client kingdoms to prevent petty local wars;[176] the Byzantine Empire sealed its 1263 alliance with the Mongols by marrying off their respective offspring;[177] competing ancient Chinese warlords coupled up their children as a form of mutual appeasement—a practice known as *Heqin.*[178] Throughout history, it seems that everywhere no love is lost, marriage has been a proposed salve.

By the early 20th century, however, intermarriage as a diplomatic modality between warring states had largely faded. If anything, populist popularization of stories about the disfiguring and ill consequences of inbreeding—the so-called Habsburg's Jaw, for instance—helped morph the practice into a caricature and a cautionary tale.

There are larger forces at play in this regard. Selection favors evolutionary efficiency, which selects for superorganisms that ultimately arborize their lineages

174. See https://www.medievalists.net/2013/01/the-use-of-gunpow.der-weapons-in-the-wars-of-the-roses/; https://en.wikipedia.org/wiki/Wars_of_the_Roses.
175. Cohen, R., & Vestbrook, R. (2000). *Amarna Diplomacy: The Beginnings of International Relations*. Baltimore: The Johns Hopkins University Press. See https://en.wikipedia.org/wiki/Royal_intermarriage#cite_note-1.
176. Salisbury, J. E. (2001). *Encyclopedia of Women in the Ancient World*. Oxford: ABC-CLIO Inc.
177. Ostrogorsky, G. (1969). *History of the Byzantine State*. New Brunswick, Canada: Rutgers University Press.
178. See https://en.wikipedia.org/wiki/Heqin.

into competing branches. Deleterious homozygous diseases represent biologic programs that breed outbreeding, and emergent stigmas attached to inbreeding could be seen as the social reinforcements of those instincts.

In other words, mutually aligning kin skin in the game as a peace strategy among warring factions today would go against the grain of prevailing cultural-biological forces. To some extent, the actual effectiveness of the strategy has been questioned in some quarters. Either way, the debate has quieted in the context of the decline of dynastic empires. That said, their failure to contribute to enduring peace may relate to a fundamental issue larger than the institution of intermarriage itself: the lack of inclusive stakeholding in running the state's affairs.

To that point, interdependence within and among competing states can now be established through many types of social, economic, and technological joint ventures, rather than through intermarriage. Peaceful habits can also be nurtured by framing battles against common threats (e.g., environmental degradation, pandemics, etc.) instead of against each other, and through the pursuit of common opportunities (e.g., space, medicine, etc.)—any domain in which stakes are broadly held.

Finally, even as marriage rates plummet around the world, rising global mobility is making intermarriage among disparate demographic groups far more prevalent today. This alone helps offset rivalrous tribalism, to a degree.

In that sense, however, the recent rise of new forms of tribalism, nationalism, nativism, and protectionism portends a more war-like period ahead. While policies and memetic algorithms that divorce groups from each other may offer internal comforts, they also undermine opportunities to interlock interests among potentially hostile factions.

If the latter trend prevails, then the separate peace, bestowed by the interconnected age, may prove to be no peace at all. Therefore, in the final analysis, there is an exigency to innovate sociopolitical, economic, and biological algorithms that foster greater interdependence and eusociality across populations.

It is striking that nature sometimes relies on evolutionary debranching to be the olive branch that promotes eusociality. That is essentially what is happening when formerly competitive individual bacteria form collaborative biofilms. Similarly, slime molds under separate duress recruit each other to form chimeric fruiting bodies.[179] Some of the dramatic and disruptive social dynamics that emerge during these examples of a reversal from solitarity to solidarity may not translate well on a human scale. Yet humans, in theory, are also in a unique position to manage this transition better—and with more agency—than any other species in evolutionary history.

Now *that* would be something worth fighting for.

179. See https://en.wikipedia.org/wiki/Dictyostelium; https://www.ncbi.nlm.nih.gov/pmc/articles/PMC2665839/.

Second-Oldest Profession

Editor's Note: This essay picks on the already battered—lawyers.

Evolution survived billions of years, on its own, without consulting a lawyer. Why then did this professional class emerge?

Looking back, for lawyers to have emerged, there first had to be laws that people couldn't interpret. That process is thought to date at least to the time of the Code of Hammurabi, a legal system invented by the despot in 1754 BC as a way to keep the mob in check. It seems unlikely that the people would even have been able to read the cuneiform script that was inscribed on clay tablets, let alone interpret it—and that may have been the point. To wit, after more than a hundred years of linguistic scholars chipping away at the code after its original discovery by archeologists, no one has been able to fully interpret the stonehead inscription based on counting fingers and knuckles.[180]

Indeed, to make cuneiform script less unintelligible, it was later translated into many other languages, including hieroglyphics—a word commonly used today to describe the illegible handwriting of physicians. The word "illegible" is derived from the Latin root "leg," which means "to read." It was all but inevitable that the legal profession, which is based on the same root, would emerge.

The utter disregard for penmanship is perhaps one of the reasons why children who visit the British Museum seem drawn to the Code's cuneiform script "with a kind of overlooked homing instinct, and they often consider clay homework in spikey wedges much more exciting than exercises in biro on paper."[181] Perhaps not. A more likely explanation is that, before they reach puberty, children can fluently decode languages that adults think are hieroglyphics. Or, maybe today's

180. See https://www.storyofmathematics.com/sumerian.html.
181. See https://www.historyextra.com/period/ancient-egypt/cuneiform-6-things-you-probably-didnt-know-about-the-worlds-oldest-writing-system/#:~:text=Cuneiform%20is%20an%20ancient%20writing,even%20earlier%20than%20Egyptian%20hieroglyphics.

museum-going junior linguists see the script in a way their parents fail to—as please-send-help smoke signals sent across time by their ancient indentured counterparts. Of note, many of the surviving tablets in the museum are thought to have been sketchily etched by the schoolchildren of ancient cultures as part of their indoctrination.[182]

As with helicopter parenting, the legal systems of paternal states are grounded in good intentions. Like the dreaded stick and the even more dreaded carrot stick offered to children, the Code of Hammurabi is a system of incentives and disincentives to keep self-interest in check. In the post-eusocial world of dissociality, transactions among competing counterparties were the primary modality of human interaction. That was bound to cause some issues. Not surprisingly, about half of the edicts depicted in the Code—thought to total 282, but no one knows for sure—are thought to adjudicate contract disputes.[183]

Which brings us to the U.S. Code. Societies built on the notion of personal freedom tend to exhibit the abuse of freedom by a few. Rules and laws governing everyone will then be generated in response. Over time, individuals' continual testing of boundaries leads to breaches that lead to more laws. This helps explain why the United States, a nation founded on libertarian principles, has ironically ended up with so many laws that the Library of Congress has stated that the true number of laws in the U.S. is unknown.[184] Once again, we see self-interest coming around to cannibalize itself.

Meanwhile, the momentum of that merry-go-round is what powers the expansion of the second-oldest profession. The more laws written, the more lawyers the system produces, some of whom go on to Congress to write more laws. This perpetual motion is evidence that the second law of thermodynamics was not written by lawyers.

182. See https://www.historyextra.com/period/ancient-egypt/cuneiform-6-things-you-probably-didnt-know-about-the-worlds-oldest-writing-system/#:~:text=Cuneiform%20is%20an%20ancient%20writing,even%20earlier%20than%20Egyptian%20hieroglyphics.
183. "The Discovery of Hammurabi's Code"; "The PLEA: Hammurabi's Code"; https://en.wikipedia.org/wiki/Code_of_Hammurabi#cite_note-12.
184. See https://blogs.loc.gov/law/2013/03/frequent-reference-question-how-many-federal-laws-are-there/; https://www.wsj.com/articles/SB10001424052702304319804576389601079728920.

But this is no reason to throw the entire profession under the bus. Dissocial societies need laws and good lawyers, of which there are plenty. The purpose of tort is to protect the public from the incentives of self-interested individuals that could harm others. The court is where self-interested views compete, and this is where lawyers thrive by the hour.

But profiting by the hour in someone's hour of darkness is not the main reason attorneys are often held in contempt by the court of public opinion. What makes the public's blood boil are the untoward externalities created when the legal profession defines stakeholding too narrowly. The fiduciary duty of an attorney is to serve the interests of the client, and the clients who hire them wouldn't want it any other way. This arrangement, however, can sometimes solve one problem while creating another.

In narrow stakeholding, an excluded stakeholder can be harmed: for bygone imperialists, it was the natives, and for today's tech empires, it's the digital natives. Adversarial process in a contested divorce where parties are represented by lawyers can lead to excluded stakeholders such as children even though the parties and the court try to serve the best interests of the children.

And, to that point, while blood may be thicker than water, to a shark—one of evolution's oldest creatures—it's blood in the water all the same.

Alumni Office

Receiving a letter from the alumni office of one's alma mater ought to feel amazing. Perhaps it does. But too often it is instead amazingly deflating, and not just because it's a reminder of an earlier, glorious age.

In an earlier, glorious age of Rome, the Latin term *alma mater* meant "generous mother." The term's first use is attributed to Lucretius' *De rerum natura,* and it became an epithet for earth goddesses such as Ceres, Cybele, and Venus. After the fall of Rome and the birth of Christianity, it lived on as a liturgical phrase describing the Virgin Mary. For example, *Alma Redemptoris Mater* is a Marian hymn authored by Benedictine monk Hermann of Reichenau (1013-1051), which is still recited today.[185] During this reincarnation of Rome's heyday, known as the Holy Roman Empire, the term "alma mater" became part of the insignia of University of Bologna—the world's first university, founded in 1088.

The first use of the term to mean one's former university or school is thought to be in 1710, and the word "alumnus" to describe the graduates of a school was born during the same period.[186] The bonding nature of the formative years of schooling and the innate yearning to belong to a tribe fostered a sense of affiliation among such alumni, including the formation of clubs. While the history of the early years of these clubs is generally fuzzy, Yale University claims the longest record of alumni activity in the United States.[187] The proliferation of such activities and associations—including those administered by the universities themselves—attests to their value to alumni.

Somewhere along the way, however, the functions of university alumni offices became, at least unofficially, co-mingled with those of fundraising functions, better known by euphemisms such as "advancement" and "development." While

185. Raab, C., & Hagan, H. (2007). *The Tradition of Catholic Prayer.* Collegeville, MN: Liturgical Press, p. 234.
186. See https://www.etymonline.com/word/alumnus.
187. See https://alumni.yale.edu/your-alumni-association/what-we-do.

alumni's willingness to support their alma mater is strong, the perception that calls from the alumni office are largely a solicitation created concern that the relationship had turned transactional: had the alumni office become a commodified version of itself?

Today, using social media, alumni are able to find each other and form groups more easily without help from their educational institutions' alumni offices. As such, the roles and opportunities of alumni offices are changing.

The most important observation is this: people can change their membership in virtually all types of groups they belong to—families, employers, and eye color—but they cannot change their alma mater. Being an alumnus of a school is an immutable lifetime membership and, with people living longer, the relationship between alumni and the alma mater has more potential value than at any other time in history.

With that in mind, we may be entering a golden age of alumni offices, one that decommodifies their mission and returns to their original purpose of serving their alumni. As the relationship shifts from a transactional model to one that promotes strong, lasting bonds between a school and its alumni, the latter's service on behalf of their alma mater will grow naturally as a result of that stronger bond.

Domestication of the Environmental Movement

The brain is a magnificent organ. On the other hand, it is also our Achilles' heel, given its vulnerability to subversion by external forces, including foreign governments and industrial interests. The feel-good hurrahs aside, one wonders if today's consciousness movements are providing the neuromuscular output of an undermined mind. Environmentalism—recycling in particular—is an example of a movement that appears to be under the industry's spell.

It doesn't take an army of neuroscientists to misdirect public brainpower on these matters. Like carnivals looking for marks, it just requires natural human inclinations and the free markets. The industry's aim is simple: create new cultural institutions where people can direct their attention, far away from the real issues, which are the relentless market forces driving plastics consumption and production.

A prime example of misdirection is the 1971 "crying Indian" advertisement, an attention-deflection campaign promoting the idea that people's littering was the main cause of environmental degradation.[188] Few knew that the ad campaign was funded by a nonprofit called Keep America Beautiful, a wolf in grandma's polyester Sunday best. The nonprofit was, in fact, funded by the polyester industry and other "American pie" institutions such as Phillip Morris and Coca Cola. Speaking of pies, these are the same tactics the food industry later adopted to persuade the public that a lack of exercise was the culprit in the growing obesity epidemic.[189] The industry took an ounce of truth and added pounds of flesh to unsuspecting American people until they became too big not to fail.

The best way to warm up to the truth about plastics is to set aside the public rhetoric and look at the cold numbers. The plastics manufacturing industry is a $600 billion a year business that contributes to a $1.2 trillion a year business in plastic products. Yet, despite being far larger than Big Pharma (annual revenue

188. See https://www.plasticpollutioncoalition.org/blog/2017/10/26/a-beautiful-if-evil-strategy.
189. See https://www.yahoo.com/lifestyle/amid-obesity-epidemic-coke-shifts-health-focus-exercise-calories-172408470.html.

$300 billion), the deflection of attention has allowed the plastics industry to avoid being labelled Big Plastics.

Alarmingly, the demand for plastic is expected to grow by an additional 40 percent over the next decade. The American Chemistry Council estimates that, to meet rising demand, $186 billion is being invested in 318 new projects to increase plastic production.[190] The excitement and new investment going into plastics production are hardly signs of a public that is truly willing to consume less and an industry that is willing to produce less.

Meanwhile, the industry domesticates the public's growing environmental consciousness and conscientiousness, not by countering the public's stewardship but by leveraging it. Recycling serves the industry's interests in at least two important ways.

First, recycling partially defuses people's anger by giving them a false sense that they have done their part for the environmental movement. From the industry's perspective, giving the public this illusion of control is essential, since feeling a lack of control inspires people to grab torches and pitchforks.

Second, the growth of interest in recycling has allowed the plastics industry not only to quell public anger but to give the industry another method in their madness. Recycling is a way for the plastics industry to socially engineer a free public labor force that increases the industry's profitability. An old saying warns, "Don't bite the hand that feeds you." Well, don't feed the mouth that bites you.

There's more. Recycling has, from scratch, created a $200 billion a year industry that has been accused of misdeeds.[191] As an industry that employs 500,000 people in the U.S., it may already have become too big to fail, and it will most likely keep getting fed.[192]

Trusting this industry to find the solution—biodegradables, etc.—or to

190. See https://www.theguardian.com/environment/2017/dec/26/180bn-investment-in-plastic-factories-feeds-global-packaging-binge.
191. See https://bir.org/industry/; https://www.theguardian.com/environment/2018/oct/18/uk-recycling-industry-under-investigation-for-and-corruption.
192. See https://www.closedlooppartners.com/3-reasons-why-recycling-is-good-business-in-america-and-a-key-driver-of-the-circular-economy/.

police itself is hardly an ideal setup. If anything, the race-to-the-bottom nature of misaligned interests points the tip of the spear—the recycling market—at the worst possible outcome for the recycled plastic: more than half of the world's recycled plastic used to end up in China, but they stopped buying it due to its environmental impact. Now, a lot of recycled plastic ends up in landfills or is burned—good intentions literally going up in smoke.[193]

Without embracing the principles of inclusive stakeholding to internalize these types of externalities, self-expanding beasts can corner the public into double binds like this: if you don't recycle, you contribute to environmental degradation, and if you do recycle, you indirectly contribute to environmental degradation. It's not unlike what happens when 100 million viable possibilities are winnowed down into two damned-if-you-do and damned-if-you-don't presidential candidates.

However, as Yossarian found in Joseph Heller's *Catch-22,* perhaps a more permanent solution to the seemingly intractable problem is hiding in plain sight. It may be as easy as walking away from the actual problem—which in this case means walking away from buying into the culture of consumerism.

Yet that may be the hardest thing to convince a brain to buy into.

193. See https://www.nationalgeographic.com/environment/2018/10/5-recycling-myths-busted-plastic/

Baseball or Moneyball?

For a significant part of the 20th century, Major League Baseball captivated the nation's imagination. Local teams were the pride, and in some ways the identity, of the town. Players often played their entire careers for a single team and were fixtures in their town's civic life. Fans' loyalty to their local team verged on the reverential. The fate of a team—think of the Chicago Cubs, who blundered their way through nine decades of championless seasons and into the hearts of professional sports lovers—not only bonded families, generations, and communities but strangers in bars and across the national landscape.

In the 21st century, baseball's role in communities has changed. To be fair, the change has come in the context of many changes in the game and in the country at large. The country is wealthier, yet the sport has priced out many casual fans from spending a season with the team. Fan loyalty is not what it used to be, due in large part to the increasing variety of entertainments available. It is also partly due to people relocating around the country for various reasons much more often than they did in the past.

After Curt Flood challenged the reserve clause rule and opened the floodgates of free agency, players also began to relocate around the country at a dizzying rate. It is not uncommon for a team's opening-day roster to turn over 80 percent every five years. Since a player's identity is rarely tied to a town for more than a brief stint, fan adoration doesn't last much longer than a summer fling.

However, you would not sense this by looking at the numbers. Baseball attendance is near an all-time high, as are revenues. But ask one hundred kids in San Francisco today to name the starting lineup of their local team—one of the most storied franchises, which plays in a beautiful stadium that is packed nightly and wins championships with regularity—they probably could not name more than five players. That degree of apathy would have been unthinkable fifty years ago.

Let's go back to the evolutionary origins of sports. Ritualized, non-lethal competition is common among social species, and human social systems exhibit ritualized competitions that result in rank. Sorting rank through competition enables the development of dominance hierarchies that feed into secondary social behaviors, including rank-based mating, resource distribution, and cooperation. Evolution favors the emergence of such behaviors if they promote fitness from a multilevel selection perspective.

The evolutionary fitness value of physical prowess is probably less relevant in today's era, when humans gather resources more through proxies such as business success rather than hunting and gathering, but the ritualized persistence of legacy behaviors is not uncommon in nature. It shouldn't be surprising that using success in sports to inform mating preference and resource distribution persists to this day.

But there's a deeper reason we are wired to give our attention to sports: a tribal sense of affiliation. Given the change that matters most—that those providing and participating in a sporting spectacle are no longer those who have kin skin in the game—the entire affair is becoming increasingly commoditized.

In the tribal era, our affiliation with competitors was likely high, given the kin skin in the game. In the Roman Republic, however, fans' affiliation with the competitors was weak and competitions were attended more for the sense of spectacle than affiliation. In the Greek Republic, something different happened. The advent of the Olympics enabled fans to develop a sense of collective affiliation to teams and athletes based on characteristics other than kinship. Uniforms, insignias, and other markers of a shared identity were manufactured to replace relatedness as a basis for feeling kinship with a team.

In 19th-century America, a sense of collective affiliation began to develop toward teams and athletes based on what city one lived in and what university one attended. Thus, the modern age of sports was spawned. As citizens grappled with an increasing sense of alienation and decreasing sense of community, due to the replacement of high-aligned kin tribes with low-aligned communities of genetic strangers, sporting leagues were more than happy to provide an emotional salve of affiliation, however fleeting. It is perhaps not surprising that sports took off as a paid spectacle in the United States, a melting pot where people felt distanced from their kin origins. They desperately needed to find common ground somewhere, somehow.

By the late 20th century, the success of collegiate and professional sports had allowed them to become major cultural institutions. A sense of affiliation among fans grew to the point of their becoming devotees who spend a disproportionate amount of time, attention, and resources on their sport of choice.

However, these sports' success at garnering attention and money led them to evolve ever more toward a commodification model. The relationship between players and fans became transactional as the industry treated it as something to be exploited for economic gain. Loyalty became a commodity to be sold and bartered. Seemingly every single part of the institution is for sale, from human bodies to stadium naming rights.

A sense of soullessness is usually a good barometer of whether an institution has turned into a low-alignment commodity. Instead of building a true sense of affiliation between athletes and fans, the business of sports is depleting the intangible values of a bygone era to maximize short-term gains, while they last. Team owners will continue to flip ownership at ever higher prices until the music stops.

The declining sense of affiliation fans have with sports teams is evident to most people today. It has been replaced with a sense of the inevitability of the decline of the role collegiate and professional spectator sports play in people's lives. As with most institutions today, cynicism about sports is on the rise.

That is not to say spectator sports is about to collapse. There is still plenty of value in the lore of the past and the roar of the crowd. Moreover, while in some ways the very soul of these games is on the line, our love of sports transcends even the games themselves.

Other Status-Change Industries

There's a deeper story about why all competitions, not just sports, draw participants and spectators. Humans are innately wired to attune not only to status but to status *changes*. If a status change is occurring within a social group, it has potential fitness import for the witnessing parties. People can participate in contrived activities for the sake of amusement, where they experience status changes themselves or witness status changes in others. Such activities are the lure of games or sports, and participants may describe their experience as "fun" or "exciting," particularly when it involves status elevation—also known as winning.

Our natural attentiveness to status changes has been exploited by the modern entertainment industry. A sport where the score is kept and a winner is declared is operating through status changes among participants. A football game might be called "great" if there are many lead changes (status changes) and "dramatic" if a status change occurs against all odds late in a game. A game where the underdog defeats the favored team might be perceived as "exciting" because of the unexpected status change. Indeed, we as consumers often demand that rank order be clarified for our own satisfaction and are willing to pay for that satisfaction.

Some people are addicted to checking score updates for their favorite teams. Others wake up on Monday morning and immediately check how the top 25 college football rankings have changed since the previous week. Casey Kasem had a long-running weekly radio show on which the top 40 song rankings were updated. The most titillating moment of the hit television show American Idol occurs when a singer, whom the audience has been led to believe is of lower status, undergoes a status transformation while performing a single song. Downward status movement, such as when a celebrity hits a rough patch, also sells newspapers to a hungry public. The gossip industry is a status-change industry.

The human mind seems voyeuristically drawn to undulating story lines of status change in books, plays, and movies. The mood changes in the scenes of Shakespeare's *Romeo and Juliet* can be described as alternating between low and high status: street brawl, intervention by the prince, Romeo mopes, meets Juliet, learns she's a Capulet, etc. Status change is considered a tool not only for drama or tragedy but for improvisational comedy, too.[194]

Facebook is many things to many people, but in many ways a Facebook page is a signaling device for status (or at least the commodification of information about status). How many friends we have, who our friends are, and what cool thing we did over the weekend are all status attributes. Every posting on Facebook changes our status relative to that of others.

Indeed, the most powerful place in all of social media today is a box near the top of our Facebook page that is labeled simply "status update."

194. See https://en.wikipedia.org/wiki/Keith_Johnstone.

On Trade

As trade uncertainties whiplash the markets like a Laurel and Hardy slapstick routine, let us take the long view on tariffs. Current tariffs are under the purview of the International Emergency Economic Powers Act, which was enacted in 1977 to give the president broad authority to regulate commerce after declaring a national emergency.[195] Ironically, the Act was intended to *restrict* the even broader unilateral powers that presidents—under the Trading with the Enemy Act of 1917—had been able to wield, providing they first declared a national emergency.[196]

Naturally, presidents have been pulling that alarm at an alarming rate.[197] Clinton declared seventeen alarms, George W. Bush twelve, Obama thirteen, and the current president has already declared many. Since 1976, U.S. presidents have declared fifty-nine national emergencies, thirty-one of which are—unbeknownst to most Americans—still ongoing.[198] While Americans might not be able to name who called which code blue when, they will do as they are told, even if it means standing around in their socks at airports.

One might laugh that off as slapstick if the following weren't also true: the power to declare national emergencies today was authorized by the National Emergencies Act, which was enacted in 1976 to stop open-ended national emergencies.[199] That 1976 law wiped the slate clean of a laundry list of open-ended national emergencies that had accumulated and been collecting dust over decades, including a 1933 national emergency President Roosevelt declared to create a bank holiday in order to prevent the hoarding of gold.[200]

The 1973 Senate special report that rediscovered this long-forgotten, cobwebbed national emergency in the legislative attic also noted that, "after only 38 minutes debate, the House passed the administration's [1933] banking bill,

195. See https://www.treasury.gov/resource-center/sanctions/Documents/ieepa.pdf.
196. See https://en.wikipedia.org/wiki/International_Emergency_Economic_Powers_Act#History.
197. See https://en.wikipedia.org/wiki/National_Emergencies_Act#Invocations.
198. See https://abcnews.go.com/Politics/list-31-national-emergencies-effect-years/story?id=60294693.
199. See http://uscode.house.gov/view.xhtml?path=/prelim@title50/chapter34&edition=prelim.
200. See https://www.presidency.ucsb.edu/documents/proclamation-2039-declaring-bank-holiday.

sight unseen."[201] That national emergency would remain unseen for another forty years. At long last, in 1976, with one twist of the wrist, President Ford reset the number of lingering national emergencies to zero, only to start the new pileup of national emergencies that we have today. It's a small miracle that our autonomous-driving government has not crashed more often.

Indeed, the U.S. government did almost crash coming off the assembly line and was saved, in fact, by a new tariff policy. After the United States successfully declared independence from the British Crown, the original Articles of Confederation of 1781 left the U.S. federal government powerless to collect taxes from each state—a near-fatal flaw for a federal government.[202] Put another way, the nation that had won the war against taxation without representation found itself stuck with representation without taxation. Needing a quick way to raise revenue that could save the government, President George Washington signed the Tariff Act of 1789, which imposed tariffs on nearly all imports and was to be enforced by the border patrol.[203] Alexander Hamilton, whose devout protectionism went unmentioned in the theater version of his biography, introduced the term "infant industries" to shape George Washington's belief that economic independence through tariffs was vital to America's political independence. By 1820, America's average tariff was up to 40 percent. From 1871 to 1913, the average U.S. tariff on dutiable imports never fell below 38 percent.[204]

The larger point is this. The most used catchphrase in Laurel and Hardy films was, "Well, here's another nice mess you've gotten me into!" It could easily have been the refrain George Washington uttered when he was unable to fund his new government, or it could be the chorus describing today's myriad issues. It certainly provided an apt soundtrack for the challenging decade of the 1930s, which coincided with Laurel and Hardy's heyday. Like now, populism, socialism, sectionalism, nativism, nationalism, protectionism, and pretty much every other "ism" was on the rise then too.

201. See https://archive.org/stream/senate-report-93-549/senate-report-93-549_djvu.txt.
202. See https://en.wikipedia.org/wiki/Tariff_in_United_States_history#Historical_trends.
203. See https://www.jstor.org/stable/1819831?seq=1#metadata_info_tab_contents.
204. See https://en.wikipedia.org/wiki/Tariff_in_United_States_history#Historical_trends.

If the current policy chaos on global trade seems disturbing to you, it's partly because nationalism and protectionism represent third-order Kayfabe reactions to the second-order issues of extractive governance and extractive capitalism, which themselves are symptoms of the first-order exclusive stakeholding that emerged when kin skin in the game did not scale during globalization.

That is to say, given that free markets in the context of self-interest produce dissocial outcomes, free trade merely amplifies this distortion. Protectionism is hardly the better model. The common weakness of free trade and protectionism is that, in both cases, self-interest serves the power structures at the expense of the people. Without inclusive stakeholding to mitigate self-dealing, free trade and protectionism are both doomed to fall well short of their stated ideals.

Heads I Win, Tails You Lose

Editor's Note: This essay describes one way the venerable exploit the vulnerable.

If you want to discuss the industry in which misaligned incentives have had the most disastrous economic impact in recent years, look no further than where you are right now. Chances are good that the building you are in was affected by the 2007-2009 meltdown of the global real estate market.

As housing bubbles form, mortgage lending drifts toward risky loans, such as subprime lending, at exactly the least opportune time. Too often, lenders try to squeeze out the last of the profits of a boom period, knowing they will only partially bear the consequences if their loans default.

Indeed, after originating the loans, they can securitize them and sell them to secondary mortgage buyers at a small profit. The availability of these secondary markets ends up encouraging transactional behaviors. Buyers on the secondary market also hope to earn a small return, despite the risk the security carries. In many cases, they too flip loans like a hot potato. When the music stops (e.g., the housing market crashes) and defaults come into play, the people who made the decision to buy the mortgage-backed securities can walk away without paying much of a personal price, even though they profited from their decisions along the way.

Some of the losses from the housing crisis were borne by the banks and their shareholders. The bank executives pocketed the gains on the way up, and on the way down, the losses were disproportionately borne by shareholders, borrowers, and the public. Moreover, the shareholders' decision to invest in banks often was made by investment managers who made money when bank share prices went up, and their clients bore the losses when share prices fell.

Taxpayers were playing the same game. From a buyer's perspective, the risks and return for home buyers are partially decoupled in a way that promotes risk-taking. When housing prices rise, homeowners with mortgages keep the equity gains. Due to the non-recourse nature of many loans, when prices fall, homeowners can walk away and offload part or all of the losses onto the mortgage holders. During the housing bust of 2007-2009 and the subsequent rounds of government bailouts, many of the losses were passed onto Freddie Mac, Fannie Mae, and other financial institutions that had to be bailed out by the government. In other words, the homebuying public was partially able to socialize their losses.

It seems that individuals and corporations are able to benefit privately from profits and to push some of the losses onto others. This ability to at least partially offset the losses on the downside incentivizes depositors, homebuyers, money managers, banks, and corporations—in effect, every stakeholder—to take excess risk while chasing yield. The decoupling of risk from return invalidates many economic models, including the concept of the efficient frontier in modern portfolio theory, and forms a rather rational basis of irrational exuberance.[205] When the music stops, everyone passes the blame to someone else.

Everyone is pretty sure somebody else was responsible for the trillion-dollar real estate bust. No one ever admitted, "I did it." Not for a single buck of it.

205. See https://en.wikipedia.org/wiki/Modern_portfolio_theory; https://en.wikipedia.org/wiki/Efficient_frontier. The efficient frontier of modern portfolio theory relied on the assumption that investors make rational decisions according to expected risk and expected return. However, when people are able to capture the upside of an investment but socialize the losses, they tend to take excess risk in exchange for higher personal returns.

Real Estate

Editor's Note: Most agents, including real estate agents, do the right thing; that said, this essay brings the principal-agent problem home to a familiar market.

The principal-agent risk in a real estate transaction is an example of misaligned incentives. While there are industry regulations, government bodies that enforce licensing requirements, and education to mitigate the risk of a principal-agent problem, conflicts of interest still exist throughout the real estate industry.

To begin with the basics, a typical residential real estate transaction involves both a listing agent, who has a contract with the owner of the property up for sale, and a selling agent, who represents the buyer and negotiates on their behalf.[206] The agents, and their brokerage houses, split a sales commission, usually 6 percent of the home sale price.

At first glance, the interests of the seller-owner and the listing agent are aligned. The listing agent earns a percentage of the sale price as a fee. Looking deeper, however, trying to get the highest possible price for a home does more than increase the agent's fee: it increases the time the property sits on the market and the probability that it may not sell at all.

For the listing agent, time is a valuable commodity, and seeking the highest possible price can reduce their overall sales productivity. Indeed, listing agents could net more income by increasing their sales volume, even if that meant selling their listings at a lower price. This business reality conflicts with the interests of the seller-owner.

Many agents value their reputation and would not undersell the owner's interests for the sake of a quick sale. A good reputation also puts them in a position to earn repeat business and referrals.

206. The selling agent is colloquially referred to as the "buying agent." Go figure.

But, the long-term relationship between agents and clients in residential real estate is somewhat tenuous. A typical client may be involved in a handful of transactions over a lifetime, many of which will involve relocating outside the area of an agent's expertise. From a client's perspective, being loyal to a particular agent may or may not pay off.

More telling is how brokerages reward listing and selling agents. Treasured metrics in the industry are sales volume and total dollars. The primary barrier to growing these metrics is the bid-ask spread between the homebuyer and the seller. Thus, a latent motivation of the listing agent is to lower the seller's price expectation, and a latent motivation of the selling agent (a.k.a. buying agent) is to raise the buyer's price expectation.

All of this is well and good, except that the fiduciary obligation of a listing agent is generally to sell the property for the owner at the highest possible price under desirable terms, while accounting for the seller's tolerance for time their property lingers on the market. The fiduciary duty of the selling agent (a.k.a. the buying agent) generally is to acquire the desired property at the lowest possible price on desirable terms.

On the one hand, agents can defend their actions to narrow the bid-ask spread. The seller wants to sell and the buyer wants to buy. But for their clients, as with all negotiated transactions, there is a tradeoff between getting the deal done and getting the deal done at the best price. The agents' economic incentive to drive that bid-ask spread down to zero can get in the way of serving the latter objective.

The frequently cozy relationships between buying and selling agents amplify this risk. Dual agency represents one of the most troubling examples of the principal-agent problem in real estate transactions. In dual agency, a single agent represents the buyer and seller of a property, which puts the fiduciary obligation to each party in direct conflict with the interests of the seller and buyer.

The cozy relationships between real estate agents and transaction vendors also pose risks to the clients. In most cases, the agents refer the vendors to the buyers and sellers. Given the lure of earning repeat referrals from agents, vendor incentives are more aligned with the referring agents than their actual clients. Thus, they may be incentivized to underplay information that would reduce the likelihood of a deal closing and overplay information that increases the likelihood the deal will close. The liability risk of this strategy is mitigated by preemptive disclaimers.

All of these incentive issues are ironically embedded in the words "real estate," whose original meaning is "to have an interest in the thing." In an inclusive stakeholding world, the "thing" would be the best interest of the client.

Remission

In September 2008, the financial markets witnessed a fortnight like no other. Lehman Brothers toppled into bankruptcy, stock prices were in freefall, and a treasury secretary was brought to bended knee trying to save a teetering global financial system—a Kayfabe moment of asking for mercy.

I spent those weeks on the road, listening, watching, and taking notes for posterity. In an age of information overload, it's unclear whether future society will remember such events, let alone care about them. Fukushima, anyone?

Observing a gargantuan interconnected system collapsing in slow motion exposed what, to me, were fatal flaws in the modern financial system (cryptocurrencies are poised to exacerbate, not mitigate, these issues).[207] I described the flaws in a series of essays, but I offered no solutions for them.[208] I couldn't think of any.

I don't think the experts could think of any either. I am a physician, and surgeons have a phrase called "peek and shriek"—slang for opening up a patient, realizing that things are far worse inside than they seemed preoperatively, and stitching the patient up without performing any procedure. When there is no cure, we focus on managing the symptoms as compassionately as possible. But let's face it; as doctors, we are just kicking the can down the road.

I believe this is what the caretakers of our global financial system did after "opening up the patient" a decade ago. The fundamental flaws in the system had gone undetected for years, when earlier detection might have offered better treatment options. Once metastasized and spread throughout the system, these issues could no longer be addressed without killing the patient. So, they just stitched the patient up and pumped them full of various fluids—a.k.a. liquidity—to prop up the numbers for a little while.

207. Yun, J. (2017, December 8). *Cryptoassets Pose Existential Threat to the Financial System, and It's Not in the Way You Might Think* (Kindle edition); https://www.amazon.com/Cryptoassets-Existential-Threat-Financial-System-ebook/dp/B078DLL4CC.

208. Yun, J. (2018, October 8). *Price and Money: Wag the Dog?* (Kindle edition); https://www.amazon.com/Price-Money-Wag-Joon-Yun-ebook/dp/B07J69FD2F/.

During the current heady times, it's easy to be unaware that we still are in this period of "a little while." Most of the world has gone back to kibitzing about who's doing what to whom and wondering what's for dinner. Such is the bliss of remission.

Money Illusion

Individuals in dissocial civilizations agglomerate as mercenaries into power structures. These power structures are inherently unstable and need to continually feed the inner beast to maintain mercenary loyalty. When the music stops, mercenary loyalty cannot be sustained and power structures begin to disintegrate. The downward spiral can take multiple institutions down and even sink empires.

Dissocial societies have no choice but to keep the music playing. The grow-or-die nature of dissocial civilizations means they are doomed to pursue growth at all costs. In the short run, restless mercenaries can be kept satiated in a number of ways. One is to increase productivity, but this cannot be sustained indefinitely. A second way is to garner gains from elsewhere. Dissocial societies are prone to pursue external imperial gambits and to cannibalize their own people through domestic colonialism. They can extract from the future through intertemporal imperialism and from the skies through interplanetary imperialism. When these gambits reach their natural limits—and they always do—dissocial societies are forced to face the music. The descent into decivilization begins as power structures unravel.

A third way—the policy of low inflation—can also temporarily soothe the restlessness of mercenaries. There are, however, major risks in this story. To some extent, the growth and inflation stories have relied on demographic expansion: the more people there are, the higher the demand for and price of resources and assets such as land or gold. But demographic growth around the world peaked fifty years ago and has been declining ever since. Perhaps not coincidentally, the U.S. went off the gold standard soon thereafter, creating an artificial means of promoting inflation.

Yet, inflation is a monetary illusion that ultimately disillusions. A run on promises—whether on our social contracts or our economic contracts—can precipitate a run on our institutions. Indeed, inflation and deflation are not self-stable phenomena, and they tend to move away from the mean. The system is non-robust on its own, thus frequent policy adjustments are needed to keep it stable. A system that is prone to booms and busts, and to inflation and deflation, is a recipe for the polarization of wealth and a risk factor for social unrest.

Overall, our financial system is one of the long-term existential risks of civilization. Whatever the final trigger may be, once growth or the illusion of money goes away, or if there is a run on promises made to various constituencies, the conditional loyalty of mercenaries can evaporate. Thereafter, the inherent instability and lack of resilience among dissocial civilizations and their associated power structures create the risk of a downward spiral.

Out Through the "In" Door

Exclusive stakeholding is a first-order cause of why large-scale systems around the world today produce perverse outcomes relative to what the system should and could produce. It's one reason why self-expanding systems start to exhibit the Byzantine reflex, the domestication, the double binds, and the race to the middle. It's why people around the globe are losing faith in their institutions.

It's also the first-order cause of why institutions and the public appear to be colluding and performing ritualized behaviors together, rather than acting in good faith—what Dr. Eric Weinstein refers to as Kayfabe.[209] We make the case that what Dr. Weinstein characterizes as the contrived nature of pro wrestling is driven by the same force that selects for high-fructose corn syrup and produces cultural commodification: exclusive stakeholding.

When a network of self-expanding, self-dealing superorganisms cooperates at an even higher level as a super-superorganism—as has been enabled by the interconnectivity of seven billion people with no mutual stakeholding—the sheer awe of the power asymmetry between the system and the people is enough to lead to mass surrender.

The resulting dehumanization is so complete that, even when you get home from the carnival, you realize you are still in it. You are still the mark of a larger game among power structures.[210] You think back to Dorothy waking up back in Kansas and find that it is you awakening to the reality that she had just woken up on a movie set. You think of Alice stepping back through the mirror right into a house of mirrors—right back into the carnival. Surely the way out is through the "in" door where we came in. But where did we come in from?

It is our contention that our best hope for finding our way out of the digestive tract of the beast is through the door of inclusive stakeholding.

209. Weinstein, E. R. (2019, October 21). "Kayfabe." *Edge;* https://www.edge.org/response-detail/11783.
210. And the power structures run the tightly-held narrative—Dr. Eric Weinstein refers to it as the Gated Institutional Narrative—that protects the system.

CHAPTER SEVEN

Story of Our Nature and Nature of Our Story

Reader's Guide to Chapter Seven

Chapter Six told stories of how we live in a highly commodified principal-agent economy. Chapter Seven tells about the principal-agent problem of the stories themselves.

"Evolutionary Dislocation of Information" outlines the genesis of the information problem—the principal-agent risk that arises when the source of information doesn't have kin skin in the game for your welfare.

On the one hand, the feedback loop of the information market selects for integrity as a social quality. On the other hand, it can also select for propaganda that exploits listeners. Gresham's law predicts that the latter will dominate the information markets.

That illegitimate cues are used to bait human attention is depicted in "Darwinian Rubbernecking." "True Story: Pulitzer Was a Predecessor to Modern Fake News" shows how this can even lead to wars.

But the real news, as suggested in "The Truth about Fake Relevance," is that fake relevance is a deeper threat to sociality than fake news. Fake relevance is harder to police than fake news since the fact checking system will perceive it as true. "Ode to Banality" recognizes that the price paid in this attention confiscation market is our blindness to ordinary but relevant cues.

Moving on from news to movies, "Concession Stand," "Story of Our Nature and

Nature of Our Story," and "The Hero's Journey as the Villain" illustrate the distortion that occurs when a market is incentivized to create stories that serve the storyteller more than the story listener.

The next two essays portray a couple of second-order symptoms of dissocial information markets. By commandeering our attention, the race-to-the-bottom free market for information naturally leads to "Attention Inequality." The escalating competition for one's voice to be heard also leads to silencing, as described in "Free Speech Eats Itself." One sign of a dissocial system is one that gate-keeps information through censorship. Another sign is a culture of self-censorship; "'Miranda Warning' for Social Media?" explores the brewing issue.

"Dissocial Media" depicts the broken promises of Internet utopia.

Evolutionary Dislocation of Information

When it comes to the information we receive from others, the change that has mattered most is who gives it to us.

We humans have an innate evolutionary tendency to trust information provided by others. Think about it. When our tribal leaders or brethren shared information back in the age of kin altruism, people could trust one thing: they did so with their kin's best interests at heart. Sure those leaders and brethren did not know much by today's terms, but when they did offer information—"There is a lion behind that tree" or "Don't drink that water"—it was in the service of their tribe's shared genetic interest. In such an environment, the human community evolved a culture of storytelling to convey information useful for survival.

Today, communications technologies and social media enable genetic strangers to control most of our information flow. Since they don't have kin skin in the game, their self-interest is to use information to exploit their audiences to maximize their own fitness. Unlike your mom, who put you first, the media industry cares first about itself and only secondarily about you. Moreover, the media industry today can dupe our moms and use them to exploit their own children.

All too often, the purpose of information nowadays is its trading value. By virtue of that commodification, the value system of the information economy can be corrupted. Given the incentive structures in place, mass media too often values velocity over veracity. They frequently present information in ways likely to maximize *their* evolutionary fitness, using information as a weapon of mass distortion. A not insignificant portion of what the media sells today is noise masquerading as information.

Even when the news is not fake, editors can fake its relevance in order to hijack our attention (fake relevance will be explored as a concept in greater depth later on in this chapter). For example, it may be true that a bear attack occurred halfway around the world, but including that story in our newsfeed distorts its relevance. With few exceptions, the media today draws our attention to a seemingly endless parade of remote threats that are not in our immediate vicinity and therefore are mostly irrelevant to our evolutionary fitness.

Like our food, the interactions encouraged by digital media are designed to be heavy on stimulation and low on nourishment. Processed information is as harmful to the mind as processed food is to the body. Moreover, the algorithms are tuned to induce attention addiction. The result is an epidemic of digitally enabled social gluttony that is every bit—or byte—as debilitating as the bites that promote the obesity epidemic. When it comes to FOMO, what's more worrisome than the fear of missing out is the fear of *maxing* out.

It is thus that we lose sight of truths, which become hidden in plain sight. Whereas a race to the bottom might select for fake news, a race to the middle might select for half-truths that blind us further. As a result, wisdoms of eusocial ages may disappear during the dissocial age, only to be relearned in the next eusocial age.[211]

211. The idea that leadership is the nurturing of the next set of leaders is an example of eusocial wisdom. That wisdom has been displaced by some who consider leadership as the garnering of Twitter followers.

Darwinian Rubbernecking

In her iconic photograph "American Girl in Italy," taken in 1951, Ruth Orkin shows men watching Ninalee Craig as she walks down a street in Florence. From an evolutionary perspective, prehistoric males who did not instinctively respond to visual cues about potential mates would face adverse evolutionary selection. In a similar fashion, when we drive past a car accident, we instinctively turn our heads to look. In nature, an organism that does not rubberneck at signs of carnage is ignoring useful cues about a threat in its vicinity, which could lead to adverse selection.

Traits shaped and culled during prior eras of evolution can be rendered maladaptive through contextual dislocation. Without kin skin in the game to deter self-interest, the media industry is incentivized to exploit our hardwired tendency to rubberneck at such cues and lures viewers with manufactured scenes involving sex and violence.

Speaking of skin in the game, in the evolutionary wild, the visual image of a nubile female in a welcoming position represents a mating signal, her willingness to incubate and nurture the male's genetic interest. However, as the case of the *Photuris* firefly reveals, predatory counterparties can mimic hardwired mating cues to their own advantage.[212] For human males today, the overwhelming majority of exposure to such visual images occurs at the hands of predatory counterparties. Porn remains the most prevalent "fake news" traversing the modern technology infrastructure in the service of exploiting human factory settings.

Our evolutionarily selected preference for social novelty is also being exploited by the social media industry. From an evolutionary standpoint, the attraction of social novelties makes sense. There are, after all, many beneficial aspects to having an increasing number of possible social transactions, whether in romantic life, friendship, or business, as access to more people can increase the probability of

212. See https://ase.tufts.edu/biology/labs/lewis/publications/documents/2008Lewis_Cratsley.pdf.

finding good partners. But, in a world where access to social novelty is virtually limitless, does a tendency to be intrigued by new social opportunities on social media make us better off?

Our innate tendency to pay attention to status—our position within the social pecking order —is also being exploited by the entertainment industry. Status can affect access to resources, mating opportunities, and many other elements of evolutionary success.[213] The entertainment industry panders to our attraction to status by hiring high-status performers and often using camera angles and staging to make performers appear taller.[214]

All of these dimensions are subject to exploitation by today's storytelling industry.

213. The luxury and glamour markets figured this out long ago. People seek constant reassurance of their status, and the consumer industry is ready to provide ever gaudier symbols of status that signal to others, and to oneself. When it comes to the luxury industry, the race to the top *is* the race to the bottom.

214. The advent of big-screen television also grabs your attention more than a smaller one because all the people portrayed on it are that much larger.

True Story: Pulitzer Was a Predecessor to Modern Fake News

One hundred years ago, the Pulitzer Prize was created to honor the best journalism. Little do most people realize that the man who financially endowed the prize had made his fortune by selling fake news. Indeed, Joseph Pulitzer invented the concept.

Fake news, a meme that emerged in recent times to call out the lack of integrity in today's information business, is descended from the "yellow journalism" movement started by the *New York World* in New York City in the late 1800s.[215] The term was coined to describe a type of journalism that treats news in an unprofessional or unethical manner, using sensationalism, exaggeration, and scandal-mongering to increase circulation rather than providing legitimate, relevant information.[216] Frank Luther Mott describes the following characteristics of yellow journalism:[217]

1. Scary headlines in huge print, often about minor news
2. Lavish use of pictures or imaginary drawings
3. Use of faked interviews, misleading headlines, pseudoscience, and a parade of false learning from so-called experts
4. Emphasis on full-color Sunday supplements, usually with comic strips
5. Dramatic sympathy with the "underdog" against the system

The race-to-the-bottom nature of the original yellow journalism era has even been cited as a cause of the United States' entry into the Spanish-American War, due to sensationalist stories printed about the horrific conditions in Cuba.[218]

215. Campbell, W. J. (2001). *Yellow Journalism: Puncturing the Myths, Defining the Legacies.* Westport, CT: Praeger, pp. 156-160.
216. Biagi, S. (2011). *Media Impact: An Introduction to Mass Media.* San Francisco: Wadsworth Publishing, p. 56.
217. Mott, F. L. (1941). *American Journalism.* New York: Macmillan, p. 539; see https:/; /en.wikipedia.org/wiki/Yellow_journalism.
218. See https://www.pri.org/stories/2016-12-08/long-and-tawdry-history-yellow-journalism-america.

The veracity of these claims is unclear, but here is an indisputable, ironic fact: the publisher of *New York World*—a predecessor of today's fake news culture—was Joseph Pulitzer.

The Truth about Fake Relevance

Concerns about fake news are rampant in the news media. Lost in the brouhaha, however, is the darker truth that fake relevance is an even greater existential risk to our information culture. That is to say, what if a piece of news is factually true but its relevance is dubious?

Take the following scenario. A train wreck occurs somewhere on planet A. Images of the wreck are broadcast to viewers on planet B, who become glued to the set while advertisers fill their subconscious minds with commercial propaganda. A third party watching this transaction might be outraged. From a utilitarian perspective, news about the wreckage on planet A is irrelevant to the people on planet B, but people's tendency to rubberneck at carnage was selected by evolution during prehistoric times, before the era when technology could send instant images over long distances. Now, our innate tendency to rubberneck at carnage, when presented by media counterparties with selfish motives, is an evolutionary lag error. The situation is akin to food companies selling sugary foods that pander to our "factory-setting" biological preferences that are wired for prehistoric times.

In the train wreck example, the undesirable impact of using fake relevance to exploit the public is self-evident. However, most uses of fake relevance in modern culture are in a far grayer area. For example, is it relevant to audience X that quarterback Y threw a touchdown pass a few minutes ago? Relevance is relative. Of all the information morsels available to consumers on any given Sunday, who is to say what is more relevant?

Sure, quarterback worship is itself an evolutionary dislocation of our innate tendency to worship our heroine aunts and hero uncles. In the bygone era, we would have shared kin skin in the game with them. Today that relationship has been commodified in service of commercial subversion. Yet is that truly exploitation, or is that just giving people what they already want? One can see how this could be argued from either side of the line of scrimmage.

But it's not all fun and games. In social structures that tend to parallax into memetic mobs, arms dealers in the information wars are incentivized to sell true truths of fake relevance (low relevance) to each side in order to reinforce their existing biases and to goad attacks against the other side in a self-escalating fashion. Can this runaway train be stopped?

It certainly won't be stopped by creating news agencies that serve true truths of true relevance. Think about it. In terms of pure relevance in the fitness landscape, we already know that the rank order of headlines on most days for most of us would look something like this: (a) not much happened; (b) most people didn't die; (c) nothing to see here. We also know that such news sites would attract zero eyeballs connected to brains that were originally wired to track carnage in the vicinity of our lives.

It is thus that the systems of today are set up to inevitably coronate the worst peddlers of bad news of questionable relevance. That is to say, even if we reran human civilization from its cradle as a simulation, free markets and natural inclinations are all it would take for media systems to self-assemble into a race to the bottom in which information is sold, not told, and serves the interest of the seller, not the consumer.

Moreover, at the point when the exploited masses revolt and set up countermeasures—e.g., fact-checking sites—the system evolves from a race to the bottom to a race to the middle. A culture of fake news gives way to a culture of true truths of fake relevance, and a culture of fake relevance gives way to a culture of quasi-relevance, recursively, until you are not even sure anymore why you are watching *The Jerry Springer Show.*

This nefarious evolutionary selection of more and more accommodative forms of hegemony—the natural survival instinct of all power structures—represents an existential threat to humanity's future. Revolt against the system itself defuses as the masses are lulled into submission by dopamine-enhancing high-truth, low-relevance memes generated by the free market of memes. Put another way, when everyone in their heart of hearts seems to want candy, you can no longer tell whether taking candy from a kid or giving it to them is the greater evil.

At the heart of the problem is the principal-agent problem. When it comes to information salience, who provided the information is more relevant than what the information was. As long as information providers do not have sufficient skin in the game for the welfare of the receiver, fake relevance will remain an intractable ailment that cannot be cured by demanding better editors or curators of relevance. Whereas the high mutual kin skin in the game of prehistoric human hives once ensured that we were informed by those who loved us most, in today's collective societies we are largely informed by those who love themselves most.

In the context of that Darwinian dislocation, here is what's germane: unless we update our bioalgorithms of inclusive fitness with social algorithms of inclusive stakeholding, we are predestined to keep accelerating toward the oncoming social train wreck ahead.

Ode to Banality

Humans are wired to pay attention to extraordinary cues. What happens, however, when "interesting" clickbait can be mass produced in order to steal your attention?

Here's what catches my attention. I am interested in things that occur in abundance. Don't stop to smell the roses—they are botanical porn. Notice instead the grass (family: *Poaceae*) growing around them. The former seduces through their preciousness, the latter commands our attention for its ubiquity, which says something about both. Why are they so successful?

While uncommonly colorful tropical fish tend to draw our gaze, most people have never even heard of the bristlemouth (family: *Gonostomatidae*), which is the world's most plentiful fish, numbering in the hundreds of trillions to quadrillions.[219] In fact, they are the most common vertebrate on earth, including land animals. Yet we know very little about them.[220] In the forefront of marine science, they are merely background noise. Who are these pervasive beings?

Speaking of land animals, ants (family: *Formicidae*) comprise almost one-quarter of the terrestrial animal biomass. However, you won't see poets swooning over their earthly dominance or professional sports franchises banking on their appeal as mascots. How did their omnipresence come to be?

Here's another category of the banal that catches my attention: systems that almost never fail. I am stunned by the flawless performance of ATMs. All I do is insert a piece of plastic into a box and punch a few keys. Yet without fail it gives me the exact amount of money I was hoping for—day after day, year after year. This kind of 100 percent fidelity of transactions has never happened before in the history of life on earth. Usually there is some asymmetry to transactions among living creatures. It's more like a slot machine or Forrest Gump's box of chocolates: you are not quite sure what you are going to get. *And that's the appeal of it,*

219. See https://en.wikipedia.org/wiki/Gonostomatidae.
220. See https://thefisheriesblog.com/2017/07/17/the-most-common-fish/.

apparently. People will voluntarily sit on a stool for hours and mind-numbingly pull the lever, like a lab rat, staring at one failed payoff after another. But people won't linger for one extra second in an ATM kiosk to marvel at perfection.

Given the number of moving parts involved, the error-proneness of the creatures involved, and the gravitational physics involved, it is stunning to think about the math of flying machines: for every commercial jet that crashes, two million don't. The former is always front-page news. The latter goes completely unnoticed: even your blood relatives won't show up to greet you after you land.

While we are on the subject of blood, unless you are a plant, you need to eat, and in the process you probably vanquish another living being. In the entire history of *Homo sapiens* evolution, we've never had so little concern about our daily survivorship. In the past, another year of your loved one not dying was not merely a I-have-to-get-them-a-Hallmark-card drag; it was a small miracle.

When it comes to human survival, vast sums of scientific funding are available to study fragile parts of our biology that break easily, such as the heart. By contrast, virtually no funding is available to study parts of our biology that rarely break, such as the hypothalamus. And all the hypothalamus (an organ the size of your thumbnail) does is regulate pretty much all the major systems in your body. Yet there is no hypothalamus interest group equivalent to the American Heart Association. The domain name for American Hypothalamus Association is still available.

Let's get this straight. When nature selects *for* the robustness of a particular function, it is probably doing so *because it is vital for survival.* But robust phenomena are selected *against* by the attention economy, including academia, *because they work.* So, while we pour money into studying the fragile parts of our body that fail us semi-regularly, we know almost nothing about the most amazing aspects of our biology that keep us alive.

You get the point. There's a lot to be learned from aspects of life that are persistent, common, and robust. They are some of the most absolutely amazing things going on. But our minds consider them boring and we pay no attention to them. In fact, we are wired for attention bias *against* the banal. Humans are wired instead to pay attention to the novel, the fragile, and the extreme. Due to evolutionary dislocations, that tendency is being hijacked by counterparties. Our collective consciousness is being increasingly crowded out by mass-manufactured extraordinary cues of fake relevance shared by those trying to seduce our attention. Today, in the post-Kayfabe world, nearly everything we consume is the bait of the extraordinary.

The paradoxical result? Ordinary is the new extraordinary.

Concession Stand

Current concerns about everyone spending too much time staring at small screens has induced nostalgia for the quaint era when everyone stared at screens larger than the largest creature to ever walk the earth. The lights were down and speaker volume was up so that no one other than the storyteller could be seen or heard (such as the couple in heat in row 27). The audience was mesmerized—eyes glued to the eye candy on the screen and shoes stuck to the shoe candy on the floor.

Regrettably, much of the would-be shoe candy would be intercepted by mouths, lured there by high-fructose corn syrup and other industrial ingredients designed by formerly underpaid PhDs to separate people from their after-tax dollars. These ingredient "hooks" were given incrementally more acceptable titles, like "Goobers," and sold at monopolistic prices at cultural economic institutions affectionately, and "confectionately," known as the concession stand.

They were never exactly clear about who was conceding what to whom. As they say, if you don't know who the mark is, it's probably you. It turns out that, as with every element of this carnival-looking-for-marks, the word "concession" is an extraction masquerading as a service to you.

All that said, boy, do we love going to the movies! They really are a nice escape—especially those hero's journeys—from the rest of our dystopian reality. It's just like the vicious cycle of Ritalin and ADHD, with each causing the existence of the other.

When you are fully immersed in the beast, that's the last thing you are aware of.

Story of Our Nature and Nature of Our Story

Humanity is trapped in a cycle.

The institutions we built to attack self-expanding beasts in one era became the self-expanding beasts of the next. We have made great strides along the way in the collective quality of life on the planet, yet there is growing dread about where the world is headed.

To step out of this cycle, we've made the case that the change that will matter most is rebuilding our institutions through the Principle of Inclusive Stakeholding, where people win as others win.

But we can do even better.

Whether through kin skin in the game or inclusive stakeholding, kindness can be induced through secondary gain—we are driven by incentives. Yet humans also display kindness for kindness' sake, even in the absence of incentives.

In the battle to overcome our default programming of following incentives—and to inspire us to show more kindness as agents of mercy rather than as mercenaries—there is a powerful secret weapon: stories.

Picture the difference between someone holding the door for another person for a tip versus someone holding the door for the sake of holding the door. The former is the commodification of kindness—done for the trading value. The latter is the decommodification of kindness—kindness for itself. One might argue that the outcome is the same either way and that only the experience differs. On the other hand, one might argue that the difference in experience is everything.

We know that people are more than capable of doing the right thing even without incentives, if they are acculturated to do so. We also know that people do the right thing even in the absence of such acculturation.

This is where stories—through acculturation—can make a difference.

The nature of our stories, however, has fundamentally changed. Over the course of human evolution, the same phenomenon that affected the evolution of human social systems affected the evolution of stories. As we became more mobile and our vested interest in each other declined, stories shifted from serving the interests of the kin hive to serving the interests of the storyteller. People began to sell—instead of tell—stories. Think of the typical Instagram post.

In the process, the storytelling market became the biggest self-expanding beast of all.

In the context of commodification, the natural dynamic of the marketplace for stories is a race to the bottom: whatever sells best at the time. The co-evolution of stories, storytellers, and story listeners selects for the high-fructose corn syrup of stories. In a way, stories themselves become self-expanding superorganisms or memetic beasts. Commodified stories are pouring in, sweeping up children and adults—who are tranquilized by "feel-good" hooks and other appealing devices—into the belly of the beast.

The story industry has optimized these hooks to maximize attention, which in itself is not evil. But the intention to use these hooks to serve their own interests at the expense of the people is concerning. Be wary of such storytellers and their industries—entertainment, publishing, podcasts, media, speaker series, etc.

But most of all, be wary of the stories themselves as they get commodified.

For example, advocacy means speaking up for others, but schools now teach *self-advocacy.* But, the more each person self-advocates, the more everyone has to yell ever louder to be the loudest chick in the nest. That's a race to the bottom that eventually becomes about the screaming itself. Victimhood and self-righteous marches ensue.

Many other words have been similarly hijacked to serve storytellers. Mindfulness used to mean being mindful of others, but now it is part of a personal-enrichment movement.[221] Alumni office means development, and develop-

221. See Yun, J. (2017). *Remarks Delivered at the 2017 Purpose Awards* (Kindle version); https://www.amazon.com/Remarks-Delivered-2017-Purpose-Awards-ebook/dp/B0763JW4YH/.

ment means fundraising. Free trade means special interests, and special interests means self-interest. Concession stand means rip-off stand, and all natural means, well, nothing.

It is our nature to nurture, but the story market picked a more marketable version of the truth as told by Thomas Paine and Jean-Jacques Rousseau, who sold stories telling us that it is our nature to be independent. Whether it's declaring independence from a king who promised to take care of us or writing a bill of individual rights, these stories address the symptoms but not the root cause of these issues—which is our abdication of responsibility to each other. On the day we took responsibility for each other, there was no reason to talk about self-reliance, a field Ralph Waldo Emerson launched in 1841.[222]

Speaking of independence, by the way, why are we still celebrating our divorce from Britain when we're now essentially living together? Shouldn't we be writing up the declaration of interdependence?

Capitalists think socialists extract and socialists think capitalists extract. And then you have Ayn Rand, who thinks all institutions extract. But who will write the true story that the lack of skin in the game is what makes them all extract? Instead, Karl Marx, Adam Smith, and Charles Dickens eloquently described cultural symptoms as if they were causes. How about the story that, with proper stakeholding, capitalism and socialism might *both* work?

But no category better epitomizes the self-expanding nature of commodified stories than the self-help movement. Imagine trying to sell a book about being mindful of others, advocating for others, and taking responsibility for others. Such stories do exist among the countless stories in the memesphere, but books that tell them (except for religious books) would find a limited audience, so authors tend not to write them and publishers tend not to publish them.

What does sell—the literary litter that self-promoting authors and publishers are willing to create—is self-help books. Next time you go to a bookstore, marvel at how large the self-help section is and how much it has grown during your lifetime. Compare it to the size of the "help-others" section of the bookstore.

222. See Emerson, R. W. (n.d.). "Self-Reliance"; https://en.wikipedia.org/wiki/Self-Reliance.

"What 'help-others' section?" you might well ask.

Exactly.

Help is a concept that used to involve a second person, until George Combe inaugurated the self-help movement in 1829.[223] Love, like compassion, is something we bestow on others, but love stories have been hijacked and mutated into stories about loving oneself and having self-compassion. These methods treat the symptoms of mutual alienation and loneliness but not the underlying causes. Who will write the truth—that the "self" movement *is* the cause of mutual alienation and loneliness? No wonder that, despite being connected to more people than ever, many of us feel more alienated and alone than ever.

Which brings us to the so-called hero's journey. One could make the case that the sum total of all the heroes' journeys in the world have gotten us to exactly where we are as a society today. What if our obsession with being the hero, slaying the beast, and being transformed is a central contributor to our current dystopia? What if our obsession with our *own* journey, heroic or otherwise, is itself a self-expanding beast of the type we described above?

People have been memetically primed to buy all of these books. The publishing industry is more than happy to profit from selling into this memetic channel, thereby expanding its own balance sheet and market power. Self-promotion becomes the engine that feeds itself through the publishing ecosystem.[224] In the competitive race to the bottom, the system selects for the high-fructose corn syrup version of every story in order to sell more books.

The "kryptonite" needed to fight these beasts is the same throughout this book: the energy of the beast against itself—the Aikido Principle. For example, one can use the incentive prize model to cure the perverse incentives.[225]

Will we truly be able to put the genie back in the bottle and all the evil back in Pandora's Jar? We are optimistic that the world is on the brink of an inclusive

223. See https://en.wikipedia.org/wiki/Self-help.

224. The system selects for fake stories with fake relevance for the reader.

225. See https://www.prnewswire.com/news-releases/yun-family-foundation-sponsors-an-incentive-prize-on-incentives----a-10-000-research-paper-prize-on-the-research-topic-inclusive-stakeholding-reimagining-incentives-to-promote-the-greater-good-in-the-new-journal-frontiers-in--300768275.html.

stakeholding transformation. However, in the free market for stories, it is easier to sell the story that this is impossible than to tell the story that it is not only possible but about to happen.

Ultimately, every yang leads to another yin and every action causes a reaction. No matter how we render the future with the best of intentions, the world will always need the next generation of stewards to shore up issues left behind by their predecessors. Inclusive stakeholding may be a salve for the panoply of extractive behaviors, but no matter how much we try to do our best, it is inevitable that a new system of victors and victims will arise. The yin and yang, the Hegelian dialectic, the path of the serpent—the shape of every form of energy—ensures that there will always be a new direction for human endeavor.

It is with this sobering perspective in mind that we offer the following thought: let's leave behind a library of uncommodified foundational stories—fairy tales of tomorrow that serve the interests of the people—that will be just as relevant into the indefinite future as they were to our primordial ancestors a long, long time ago.[226]

226. In practical terms, the storytelling industry is our best ally for transforming our current stories into their decommodified forms.

The Hero's Journey as the Villain

Selfhood implies a first-person perspective.[227] In doing so, it tautologically creates otherhood. Since self-dealing is inherent to the "self movement," we assign attractive qualities to our own ego and less attractive qualities to others.[228] The feed-forward nature of the self movement expands the gulf between self and other. Separatism is a self-expanding concept.

We argue that the separatism of the hero and anti-hero archetypes is an emergent property of the self movement. As a matter of existential convenience, people identify with the hero archetype. In doing so, they conveniently dump all the burdens on the anti-hero "beasts" they are supposed to slay.

One notable example is Joseph Campbell's idea of the "hero's journey," which he described as the story of individuals who, through great suffering, reach an experience of the eternal source and return with the means to set their society free. Typically, a person leaves home on an adventure, slays a beast, and returns home to close their journey of transformation and redemption.[229]

A couple of features of this so-called hero's journey are worth mentioning. First, in *The Hero with a Thousand Faces,* Campbell makes the case that the hero's journey is a monomyth that comprises the vast panoply of enduring stories that stretch back to ancient epics and the Bible itself.

227. Self-hood implies a first-person perspective.
228. We will later discuss our tendency to split the universe (a word that literally means "one story") in two by separating it into a black-and-white story. "That [the narrator of Genesis] transferred the impulse to temptation outside man was almost more a necessity for the story than an attempt at making evil something existing outside man"; von Rad, G. (1973). *Genesis: A Commentary* (rev. ed.). Philadelphia: The Westminster Press, pp. 87-88. Later, in the spirit of recursive duality, we discuss how original sin is the original gift. In the spirit of recursive duality, we will contradict ourselves and note that the commodified version of Genesis, which protects Man from realizing that he is part good and part bad, was a gift and not a sin. Thus, the expulsion from Eden was actually a launch. The ambiguity of this concept is both its curse and its gift.
229. In other variations, such as Dickens's "A Christmas Carol" (subtitled "Being a Ghost Story of Christmas") and Frank L. Baum's *The Wonderful Wizard of Oz,* it is not entirely clear if any physical displacement actually occurs in the course of the hero's journey. Beasts (or, in Dorothy's case, a pair of witches) are encountered and (again in Dorothy's case) slain in one or another transient, mythical land, but the true journey is internal. The hero returns home, but in fact the hero has never left. The beast that was overcome and the power to return home was within the hero all along.

The second feature is more subtle. The appealing narrative of a person whose life journey is to become the hero and be transformed as a person is another instantiation of the self movement. It is a natural extension of Freud's concept of ego and Maslow's concept of self-actualization.

One could make the case that the sum total of all the hero's journeys in the world has gotten us to the place we as a society occupy today. What if the narrative of being the hero, slaying the beast, and being transformed is a central contributor to our current dystopia? What if our focus on our own journey, heroic and otherwise, is itself a self-expanding beast?

There's probably a reason why this so-called monomyth has needed an endless series of sequels, with those on the production side commanding ever higher fees. Their enduring appeal speaks to the possible reality that the need for that story to be retold is only growing, possibly a result of its own success.

What if *we*—all of us, manifesting our most heroic "kill-the-beast" selves—are at the core of the very problem we are trying to solve?

Let's get back to fundamentals. Stories once served kin tribes. But once we shifted to low-alignment communities, storytellers' interests shifted to seeking secondary gain for themselves. Stories became commodified. The marketplace for stories started to select for commercial appeal rather than for service to the kin tribe. It is this marketplace that selected the hero's journey, with its simple and appealing archetypes: black versus white, evil versus good, beast versus hero.

The hero's journey stories are far more self-satisfying than stories that portray us as mixed characters. The former story template has crowded out the latter. Meanwhile, the apocryphal hero's journey story has spread everywhere as a self-replicating beast, the high-fructose corn syrup of stories. We are lured to both by the tip of our tongues.[230]

If the dominant stories we are hearing are the mutated ones, what might be

230. Sugar receptors are on the tip of the tongue. Stories also arise from the tip of the tongue.

the elementary ideas that once served the interests of kin hives? Is there any way to reconstruct the fairy tales of that era?

Why reconstruct these original fairy tales at all? In our view, the fairy tales of the kin-tribe era will be an important part of the cultural reprogramming needed to help complete humanity's transition from the kin-village era to the global-village era. With ten millennia of hindsight to guide us, might humanity at last be able to escape the seemingly endless game of institutional whack-a-mole in which we appear to be trapped? And if we're not to do so as modern-day heroes slaying dragons, then how might we proceed?

Moreover, if such a story exists, why hasn't it written itself? Why hasn't it been induced by demand? Think of what might have been the original fairy tale. That story is not elusive. Like many truths, it's actually hiding in plain sight. We don't need to look further.

So here it is. If we change one thing from modern fairy tales—that duality is *the* reality—the fairy tale of the tribal era emerges from hiding. Each of us is simultaneously the hero and the anti-hero. There is no distinction of subjectivity between Beowulf and Grendel. There is no separatism of black and white. We are all grey.

Imagine that the aim of our journey is to steward each other's journeys.[231] In other words, be each other's keeper. We believe this was the fairy tale of the kin-tribe era. We believe this was the original love story. We believe this was also the original hero's journey, before it transmuted into a self-expanding beast. We are never just the hero or just the anti-hero; we are a hybrid of the two. Beowulf and Grendel return home together, and there is no separate assignment of a better or worse ending to either part.

Is ice cream, a recipe for childhood diabetes, the worst evil perpetrated by Big Food? Or the best food ever, as children attest? The truth is, ice cream is just ice

231. A standard hero's journey is Beowulf, in which he slays a beast named Grendel. Instead imagine a more "authentic" hero's journey, the steward's journey. In the steward's journey, instead of naming the "other party" as Grendel, the steward names the other party as Beowulf. The steward's journey is to steward Grendel's personal transformation into Beowulf so that *he* can go home as the hero.

cream. Whether ice cream is good or bad is a matter of the observer's perspective, which means it's both. A curse is a gift. Just look at the hundred-year loyalty of Chicago Cubs fans. A gift is a curse. Just ask some winners of the lottery. Was Mother Teresa's selflessness selfish? As much as we want to disassociate, these are two sides of the same coin.

Real individuals don't embody simple archetypes in the way they are portrayed in popular good-versus-evil stories: angel versus devil, Superman versus Lex Luthor, etc. Yet these are the stories that travel far. Anakin Skywalker *is* Darth Vader. But to sell tickets, they had to be displayed as two separate characters spread out over years of sequels—as the hero *or* the villain, not the hero *and* the villain.

Meanwhile, the "original" fairy tale of the tribal era—that all of us embody the duality of being both good and evil, and that our real hero's journeys are to steward each other's journeys—has less intrinsic appeal as a story. It's a more complicated story and has less commercial value.

The human mind has trouble accepting that duality is our reality. Our brain tends to view people as either good *or* evil, instead of everyone being good *and* evil. Fervent supporters of the political left believe they are the holders of truth while the other side is evil. Fervent supporters of the political right believe the exact opposite. The story that each side might be *both* is harder for people to accept.

True truths are far less appealing in the story market than fake truths.

That's a sign of the truth.

Attention Inequality

Editor's Note: In an attention economy defined by exclusive stake-holding, power structures are colonizing and cannibalizing the weakest individuals in the system.

While the topic of rising wealth inequality has been getting a lot of attention, another area of rising disparity has gotten far less mention: attention inequality.

In the attention economy, attention is wealth. The reverse is also true: wealth attracts attention. When rising attention inequality and rising wealth inequality feed on each other, celebrity becomes a self-expanding beast.

The size of this beast can be estimated as follows. Take the Gini coefficient, which is a measure of wealth inequality, and replace wealth with a measure of attention: followers, views, citations, etc. This gives us a snapshot of attention inequality. Not that they need any more attention, but let's call this the Kardashian coefficient, since the name epitomizes the idea of being famous for being famous.

Attention inequality has been growing for a very long time. For most of human evolution, humans lived in kin tribes. The gulf between the most attention-getting and the least attention-getting members of a tribe was limited by the size of the tribe. Even the most attention-seeking among us probably had no more than a few hundred face-to-face followers.

Once humans found a way to replicate faces and names and impress them upon non-kin, attention could be hijacked to serve a larger, extractive economy. As mass media emerged, that gap widened. With transistor radios amplifying his feats, Babe Ruth garnered a heretofore unseen degree of celebrity—leading him to claim, when asked to explain why he deserved more money than President Hoover, "I had a better year."

Then there was John Lennon who, later in the 20th century, asserted that the Beatles were "more popular than Jesus now." And sure enough, his estate is still collecting his posthumous royalties. We seemed headed to a winner-take-all world of celebrity, one where you *can* take it with you.

But a funny thing happened on the way to the future. The advent of the Internet fundamentally disrupted the attention economy by giving billions of people—and their cats—their own media platforms. Elite celebrities can no longer sustain getting a disproportionate amount of attention the way the Beatles did during the halcyon days of monoculture.

Today, the attention economy at the top has become a continuously competitive race to the bottom, with ever fewer celebrities commanding the most mass attention for any significant period of time. Gangnam Style, for example, went out of style in an instant, and few remember that it even happened. Christiano Ronaldo and Ariana Grande, the two celebrities with the most Instagram followers, could walk down many streets in the world without causing a stir. As we observe the beast of celebrity consuming its own tail let's, in the wise words of Paul McCartney, "let it be."

To the broken-hearted masses, however, there's been no answer to these times of trouble. Sustained worship of monoculture celebrities has been replaced by a continuous rotation of attention from one short-lived celebrity to another.

As a result of this race-to-the-bottom clickbaiting for your attention, vast numbers of people are receiving almost no attention from others. It's likely that each person needs a minimum amount of attention to feel human, and if you were to draw the "attention poverty line" at that point, you'd see that the number of people starving for attention is exploding.

While loneliness is hard to measure, what can be quantified is attention deficit. It brings new meaning to attention deficit disorder, which is probably the fastest growing social condition in the world. No dose of Ritalin or Instagram offers a cure. Imposters—extractors masquerading as providers—are moving in to fill the void and to cash out the last vestige of human dignity.

But even in this hour of darkness, there is still a silver lining shining in the electronic cloud. As the gap widens between the haves and have-nots in the attention economy, and as the opportunity cost of attention rises, the attention you *do* give someone is now more valuable than ever. Every evening spent alone with a friend, every act of kindness to a stranger, every message you write in complete and well-crafted sentences (on paper and in email) carries more weight than at any other time in human history. Imagine what it means to be chosen over seven billion other people in that moment.

We feel that love vividly when others shine the light on us, and we spread the same joy every time we shine the light on others—reminding them that they, too, are not invisible.

Free Speech Eats Itself

The more people attack each other's opinions as a way to speak up for themselves, especially on the Internet, the more people become afraid to speak up. Free speech eats itself.

"Miranda Warning" for Social Media?

How often have you seen someone post something on social media that makes you think, "That's not going to age well"?

We live in a world where people's lives and livelihoods are being annihilated for something they once said, perhaps in a cultural context radically different from where they are now. You are probably often surprised, then, when people blithely post things on social media as if there are no potential consequences down the road.

This is particularly concerning for young people, who have long lives ahead of them but are often short on common sense.[232] Few slow down enough to consider the possibility that what they write today will be judged later by a different standard.

Thus, reformers may argue for a Miranda-type warning for social media—a "Media Miranda" along the lines of, "You have the right to remain silent. Anything you say can be used against you in the court of public opinion." These words are lifted almost verbatim from the real Miranda warning, a law—backed by the U.S. Supreme Court decision *Miranda v. Arizona* in the summer before the Summer of Love—that requires police officers to give a specific warning to people taken into custody.[233]

Despite their nearly symmetrical wording, however, the use cases of these two warnings are distinctly different. The first two sentences of the Miranda warning stem from the Fifth Amendment privilege against self-incrimination, envisioned as a protection of individual rights against the potential tyranny of government.[234] A Miranda-type warning for social media, on the other hand, would fall into the category of a consumer warning, like the Tobacco Control Act of 2009.

232. See https://pediatrics.aappublications.org/content/140/supplement_2/s67.
233. See https://supreme.justia.com/cases/federal/us/384/436/.
234. See https://www.law.cornell.edu/constitution/fifth_amendment; https://www.law.cornell.edu/wex/self-incrimination.

As a consumer warning, there would be naysayers to a Miranda-type warning for social media. Those who advocate for free speech and individual responsibility have long contested regulations that require economic institutions that wield power over individuals to alert those individuals to the potential harm of using their products or services.[235]

In addition to a potential debate about the legal logic of a Miranda-type warning for social media, there is also the question of whether such a warning would achieve the intended results or would backfire. Some have contended that the latter happens when youth are warned about the dangers of tobacco use.[236] Youth are rebellious creatures, and self-destructive behaviors are notorious status symbols that nurture their own following. How often have we seen people, especially youth, say outrageous things just to get a rise out of an audience? Such is the marketplace for attention in the race-to-the-bottom attention economy.

Lastly, scant is the political will for any measure that may or may not protect citizens from harm that may or may not be inflicted on them in the future. There is no way to prove that the tendency of information markets to use what people say against them will cause harm in the future. Indeed, the future may be far more forgiving and accommodating of things said in the past—a hat tip to a society that values legal free speech.

That said, dissenters could argue that social risks are rising in the social media age. The exact moment when people have the most power to express themselves appears to be, ironically, when institutions have the most power over them by playing the game of heads they win, tails the people lose.

What would motivate any of these institutions to ever selectively use what people once said on social media against them?

You have a right not to answer the question.

235. See https://www.bbc.com/news/world-us-canada-14553228; https://www.nydailynews.com/opinion/mcdonald-suit-happy-meal-toys-california-mom-monet-parham-new-responsible-parenting-article-1.472666.

236. See https://www.industrydocuments.ucsf.edu/tobacco/docs/#id=gmnx0096.

Dissocial Media

Editor's Note: A significant risk of Exclusive Stakeholding is the subversion of the precautionary principle, whereby innovations are thrust upon the world by its purveyors without due caution regarding the long term risks.

The West turned wild again with the birth of the Internet during what, in retrospect, were the go-go-yet-quaint 1990s. Prevailing winds infused pioneers and poets with a breezy sense of "let's go explore the world together."

By 2020, the prevailing sentiment on the Internet was "let's destroy each other." A mere twenty-five years after we began moving things to the cloud, it had darkened on us.

What happened?

Some lay the blame on mainstream media and social media institutions. They worry that the pursuit of profit promotes polarization. Like all complex situations, however, there is no singular explanation.

Yet we argue that there is no phenomenon more singularly important in human evolution than the dislocation of our social structures from eusociality to dissociality. All hell effectively broke loose when individual bioincentives shifted from "you win as others win" to self-interest. Thereafter, bands were held together by transient transactional loyalties among mercenaries rather than by loyalty fostered by kin skin in the game. After breakups, bitter dissent reigns.

Often, what begins as feel-good institutions devolve into a race to the bottom among power structures that subvert rather than serve individuals. What starts out as feel-good individualism devolves into a race to the bottom among demagogues who extract power from others to serve themselves. Eventually, malignant power structures and demagogues overreach until they are destroyed by revolutions that reboot the same everything-eats-itself power cycle.

This ouroboros loop is the dissociality supercycle.

Monarchies, republics, and even the revolutions against them grow, then die, according to this same "if-then" recursive code sequence, resulting in a self-referential fractal that looks different with each iteration of historical recursion. Seen in low resolution, the arc of history has been an inexorable race to the middle toward systemic tyranny masquerading as progress—a world in which no one even realizes they are living in a simulated existence and alienated within their own reality.

In this context, each iteration of power, technology, and communications innovation accelerates existing trends triggered by the eusociality-dissociality dislocation. That is to say, innovations such as language, numbers, horses, steam, electricity, telephone, radio, television, and the Internet speed up the rise and downfall of supercycles. Dissociality supercycles that used to take centuries (e.g., the rise and fall of the Bronze Age dissocial empires) now happen within decades.[237]

Once seen, the watermark of this historical pattern cannot be unseen. The ideals of individual liberty triggered self-interest-based capitalism, which had a seemingly good run until it was undone by its own excesses. For example, the charlatanism, power polarization, and collapse of the 1920s ushered in an era of unrest and regulation of individual freedoms. Power structures and demagogues who subverted in the name of service rose to prominence with sufficient momentum to usurp power during the 1930s.

Leaders of all types of dissocial power structures—consider every empire throughout history—find it necessary to feed the ever-expanding beast, thereby triggering both external and internal imperial gambits, in order to give the paid-off mercenaries annual raises. These unsustainable gambits drain value from the commons until their societies implode. And yet, both capitalist and socialist ideologies see the perils of self-interest as the terminal pathology of the other while failing to see their own. This self-blindness becomes the source of failure, which completes the life cycle of dissociality.[238]

237. See https://en.wikipedia.org/wiki/Bronze_Age.

238. Epochs of American cultural history reflect similar recursive supercycles. The institutional oppression and suffocating witch hunts of the sanitized-but-clean-cropped 1950s gave way to the collectively unsanitized

In the case of the Internet, we are witnessing the rise and fall of a dissociality supercycle in just a few decades. When it first came into being—the unintended consequence of each person connecting to a self-growing network—the Internet had a utopian feeling. The fling began in the 1990s, when organizations pursuing narrow stakeholder strategies burst onto the cultural landscape, dressed up—or dressed down—as dot-com clown shows. Early creeds espousing the idealistic commons of the Internet had fallen prey to a race to the bottom among charlatans chasing quick bucks, until everyone and their grandmothers were left holding a wilted tulip in their E-Trade accounts.

Then everyone and their grandmothers got social media accounts.

This was a bit like the founding of any republic, where individuals find their voice and become empowered to contest the tyranny of institutions. As in the early days of every revolution, the rhetoric and ideals of the early social media age were rosy, friendly, and utopian.

Thereafter, the rise of individualism—an emergent symptom of dissociality—created an unprecedentedly powerful race-to-the-bottom market for demagogues who extract power from others so they can use it for themselves. We came to know them as influencers. The enormous market power of influencers became the spoils that the masses chased in hopes of becoming the chased.

The gambit for followers on social media became a frenzied free market for self-interest unlike any other. The social media landscape became littered with digital mercenaries. The market sorted itself quickly into blue-check demagogues and Internet elites, while the rest were left choking on exhaust fumes, chasing whatever had already left the station. Eventually, sociopathic behaviors became the snooze that is the Internet.

Soon the operating system of dissociality ran the recursive subroutines of mob mentality and virtue signalling. The same incentive—self-promotion—that makes teen followers parrot popular TitTok dance moves also drives too

kumbuya of the don't-look-at-my-long-hair 1960s, which—after the spin cycle of the disco era—gave way to the look-at-my-hair-narcissism of the 1980s, which gave way to the don't-look-at-the-laundry-I'm-wearing 1990s, etc.

many institutional leaders to parrot prevailing views. This nefarious subroutine activated revolutionary parallax among the skeptical, who started their reactionary insurgency.

It is at this point that the entire exercise of self-interested media companies flipped from creating relatively well-meaning institutions to becoming profit-gorging beasts. They are incentivized hereafter to algorithmically turn social polarization into 50-50 battles that help the company meet the quarterly earnings expectations of shareholders. Social pollution is the excluded stakeholder of dissocial media companies.

By most accounts, the early euphoria of the Internet has turned into dysphoria. But complaining about the selfish behavior of social media giants in 2021 is like complaining about Ikea furniture on the Titanic. There has been a participatory destruction of democracy in which everyone has played a role—as is true of all forms of socially mediated self-destruction.

But what has been particularly remarkable about the Internet dissociality supercycle has been its speed and volatility. Compared to prior wild rides, this one has taken place at whiplash speed. It took place within about one-third of a typical human lifespan, which means a huge portion of the population has seen how the beast of the Internet is eating itself.

However, to wax powerless at the moment when things become what they fear is to auto-run the next line in the dehumanization script. The ouroboros of history—the tendency of systems to eat themselves—is not the root code. The tendency is a symptom of an underlying root cause: the shift from eusociality to dissociality.

The Internet, despite its failings, has delivered on many of its original transformative promises and more. While the prosocial ethos of the Internet's start-up phase may have taken some lumps, the good will flowing through the network remains enormous. If inclusive stakeholding can mitigate some of its sociopathic uses and promote more prosocial ones, perhaps we will come to learn that the utopian dreams of Internet pioneers had not been so far fetched after all.

CHAPTER EIGHT

Toward Universal Eusociality

Reader's Guide to Chapter Eight

Chapter One offered a view of the social, economic, and political ills of history as separate phenomena or as symptoms of a common root cause: the decline of human eusociality.

Thinking at the root-cause level helps us imagine cures in Chapter Eight.

"Second Coming of Year Zero" suggests that we may be near an inflection point in history.

"First Principle of Eusociality" proposes inclusive stakeholding as the underlying ab initio *principle from which solutions for the inflection point can be derived.*

"Revolutionary Thinking" suggests that we can develop solutions now by changing our mental models of existing data—as Copernicus did—rather than by gathering more data.

"Interdependence Day" is an example of rethinking our mental model of sociality. We've tried two different frameworks for the American Dream: "self-reliance," and "I rely on you"—a.k.a. the welfare state. The one that can help us break through to the other side, however, might be one we haven't tried: "You can rely on me." May these be the Oaths of Responsibilities that will go hand in hand with the Bill of Rights in a eusocial future.

"Psychology 2.0" notes that the mental model change needed is as simple as what Copernicus showed us—put someone beside ourselves at the center of our existence.

"I Know Why the Songbirds Sing" refers to the role of song in cultural transformations. "Oh My, My Captain" imagines a potential world that reveres stewards.

Consider the next five essays as a collection: they walk readers through implementation strategies for turning excluded stakeholders into included stakeholders. "Internalizing the Externalities" offers specific examples of what an imaginary industry might look like after the existing industry model is transformed through inclusive stakeholding. "Universal Basic Stakeholding" argues for a better version of universal basic income, and "Interdependent Capitalism" makes a more general economic argument. The final essay in this group, "Systems Philanthropy," discusses the importance of doing the best we can to expand our frame to include the total system.

"Can Eusociality of Our Cells Improve Our Health?" then uses the concepts explored in the book to take a look at human aging and cancer as fractal analogies for what is happening to human civilization as a superorganism.

The next four essays walk readers through suggested ways to catalyze the implementation of inclusive stakeholding. "Self-Driving Revolutions" focuses on the notion of deploying a strategy by relying on natural forces. "The Ouroboros" and "The Race to the Top" explore ways to accelerate the implementation of a strategy in the face of potential resistance by the status quo.

Finally, "Life, Itself" is an essay of hope for a future when no cultural re-engineering through reimagined incentives is even necessary. It explores the idea that, once we truly own the curative vision of a truly universal eusocial future, we can simply render it into existence.

The Second Coming of Year Zero

What will the year 2020 be called in hindsight?

The adoption of the Gregorian calendar as the timeline of history has its own notable history.[239] It is rooted in the Anno Domini system conceived by the Scythian monk Dionysius Exiguus in 525 AD, which he based on estimating the number of Easters celebrated over the years. Until that time, the unfolding of years had been named after the ruling consuls, a tradition dating to ancient Rome.

The age of ancient Rome wasn't assigned a date until the time of Marcus Terentius Varro, under the aegis of Julius Caesar, in what is now known as the year 47 BC. After compiling the tenures of Roman consuls—and the spans of "dictatorial" and "anarchic" years—Varro estimated the age of the republic, *Ab urbe condita,* to be 706.

However, Caesar could not have foreseen a future in which historians would count down the years of his rule toward zero. From his own frame of reference, time, like his armies, would have been expected to march only forward. And perhaps it would have, had he and the republic not had debts to pay to fate on the Ides of March.

239. Evolution of social systems favors the emergence of memetic interoperability: a honeybee's waggle dance, a lightning bug's flash, a nightingale's ode. Throughout their evolutionary history, humans have also developed advanced symbolic systems to coordinate communications among in-groups. As with genetic speciation, however, lineage dispersion also leads to memetic speciation until interoperability is lost. Moreover, as mutual kin skin in the game declines—as both a cause and effect of genetic dispersion over the generations—descendents of former kinsmen turn into warring factions competing for resources and supremacy. The winning faction's symbolic system tends to descend.

We live in a world that is reconverging into a global tribe after a long epoch of genetic and memetic diasporas. As earlier bloody battles among outgroups give way to cooperation and reintegration, memetic despeciation enhances interoperability. Widespread use of alphanumeric symbolic systems on the Internet is one example. The near-universal adoption of the Gregorian calendar and the 24-hour system devised by Egyptian astronomers is another.

That a common system of timekeeping was adopted faster than common language during globalization is no accident. Without proper temporal synchronization, every intersection would be an accident waiting to happen and we couldn't meet up to see *Waiting for Godot* together. With the intersection of our previously disparate time systems also came the temporal synchronization of our pasts. A prevailing view of how we got to now got placed onto the Gregorian timeline by those who prevailed in the West.

And that brings up the larger point. The post-hoc dedication of the events of this era—including the birth of a certain carpenter—as the pivot point on the ledger of trips around the sun reflects the transcendental impact of this time period on the arc of history.

The story of that arc goes something like this:

> *Inclusive Stakeholding is the first principle of eusociality, and we have outgrown nature's version of it—the kin skin in the game of our prehistoric ancestors. When kingship replaced kinship, sovereigns began ruling over, instead of on behalf of, the people. Ancient republics created to counter these abuses also collapsed, due to self-dealing. As republics and empires everywhere collapsed from corruption, a story of God who gave his only son to the world emerged to counter the story of kings who gave the world to their sons.*[240]

Thereafter, from the frame of reference of Dionysius and other observers of belief systems that predominated over Europe in the centuries that followed, human civilization must have finally appeared to be moving forward after eons of regression into depravity—in an Einsteinian relativity sort of way.

But, just as Caesar had no idea that his era would later be seen as a period of regression by revisionists, those revisionists basking in sunlight never imagined that Petrarch would later label *their* era the Dark Ages. That's because the Church had replaced the state as the latest vessel of brutish self-dealing and extractive usury—until Martin Luther led a revolt in 1517.

His revolt spawned the Second Dark Ages, also known as The Enlightenment.

And that's when the real trouble began. In the centuries that followed, self-expanding social, political, and economic power structures began taking over large swaths of the human experience.

Today, it is becoming increasingly clear that we have once again been betrayed by our institutions. But not to shrug off Ayn Rand's polemics, it is not the nature

240. See https://www.amazon.com/Essays-Inclusive-Stakeholding-Conrad-Yun/dp/1949709841/.

of institutions to fail us. We have betrayed ourselves by failing to understand our own nature—to understand that the current era is nothing more than another aftershock of a seminal evolutionary dislocation: our kin-skin-in-the-game hardwiring has scaled poorly as the operating system of human eusociality during globalization.

That's the first-order cause that continues to spill higher-order self-renewing series of dramas from Pandora's Jar. Without eusociality to protect against the unfortunate consequences of self-interest, the race-to-the-bottom tends to select for the reincarnation of dissocial power structures that rule over, instead of on behalf of, the people. We have been blind to this fundamental truth: even on the AD side of Year Zero, civilization remains one big principal-agent problem where agents occupy the superpositions.

But a brand new era is about to dawn. Starting again with the First Principle of Inclusive Stakeholding, we can build social algorithms that allow kin eusociality to be reborn as global eusociality. Whereas malalignment with competition is a race to the bottom, alignment with competition is a race to the top. If we succeed, we will return from the seemingly endless pilgrimages of history and finally all head home again—not as transformed heroes or as prodigal beasts but as brothers and sisters destined to steward an interdependent planet. Looking back in hindsight from that imagined future, we are living today, unbeknownst to us, near the end of millennia of regression, counting down to the pivot point of history when global eusociality will be launched.

It is time to spread the good news: the Second Coming of Year Zero lies not far ahead.

First Principle of Eusociality

Editor's Note: When the unit of human sociality shifted from the kin village to a global village, we outgrew nature's version of inclusive stakeholding: kin skin in the game. The future of inclusive stakeholding is the biomimicry of inclusive fitness at larger scales of sociality.

The First Principle of Eusociality is inclusive stakeholding: that is, assigning a stake to others in the widest sense, including those who currently don't have a voice, such as our future children. From this First Principle, large-scale human history can be derived and a better future for all can be imagined.

Evolution selected our social instincts to align with kin tribes. It probably was difficult for early humans *to avoid* living in kin tribes. The benefit of kin-based living was too high, as was the cost of avoiding it. That was our Eden—our social nirvana of time immemorial.

In that cradle of human evolution, there was less need for consciousness. You followed your instincts and things generally worked out. It might be no accident, then, that the kin-tribe era left us no record. Perhaps life for them just *was*.

But somewhere along the way, humans harnessed the Promethean Fire and learned to make tools. That knowledge uprooted humans from their kin tribes and the era of social entropy began. As lineages arborized, kinship thinned. The hive became a house divided. Descendants battled and were banished to a life of wandering. The dispersion and diasporas hit their planetary limits and merged into melting pots.

Looking back, the journey to now has been mostly a beautiful one. But the human experience over the past ten thousand years has also brought an ever-increasing awareness of the existence of good and bad, expressed through the service or disservice of others, either according to or in spite of Hamilton's rule.

The battle between these forces has been as dramatic as the prehistoric kin-tribe era was undramatic. Every collision between kin-based societies and societies built on competition—for example, when the "Native Americans" encountered the Europeans—resulted in the annihilation of the former due to the advanced weaponry possessed by the cultures with commodification. But over longer cycles, even the latter category of cultures built on commodification imploded. Thus, as per Ibn Khaldun's *Asabiyyah,* all civilizations began to rise and fall.[241] Strange things started to happen regularly. It began to be worthwhile to observe, contemplate, and act in profoundly new ways.

By and large, we humans have had to learn by trial and error. Ideas that seemed good to the parade of conquerors, revolutionaries, and social reformers that have disproportionately shaped our collective history delivered disappointing—if not downright destructive—results. Yet, we were never quite sure why. Even though we've made great progress in improving our material existence, the rising tide that has buoyed us materially has also unmoored us spiritually. Too many parts of the human experience feel soulless. For the vast majority of people alive today, the price of progress has been a loss of the sense of belonging that was the anchor of human experience for most of the history of our species. This cost has been significant—and, even more to the point, one we've not even begun to seriously recognize or address.

We are genetically wired for the kin-tribe era, but we no longer live as such. That's a first-order problem that continues to spawn second- and third-order issues in a degenerating cascade. So, the time has come to return to our roots. It's time to restart at year zero.

Just as Caesar had no idea that future historians would count the calendar years of his reign in a backwards regression to zero, just as every person basking in the sunlight a thousand years ago was unaware that Petrarch would later label their era the Dark Ages, and just as everyone once thought the earth was bigger than the sun, we today are blind to the possibility that our journey forward has been a journey homeward.

We left our kin village. We envisioned our role as the hero that would slay the beast. We instead got cast in the role of the villain by an unconditionally loving

241. See https://en.wikipedia.org/wiki/Asabiyyah.

hero. It turned out that these stock archetypes are the farthest things from the truth. There never was a beast to slay or a hero to worship. The same actor has played all the roles. Us.

The story we are telling is that our lives—all of our lives—are best served by encompassing some mix of both modes of living: pro-social and pro-individual. Enlightened self-interest as a social code works best when we also have alignment with and a vested interest in each other's lives. The change that *will* matter most is to update the bioalgorithms of inclusive fitness in the tribal era with the social algorithms of inclusive stakeholding in the global era.

That's because, seen through a wider lens, human history is a story of how kin skin in the game has scaled poorly as the operating algorithm of human sociality in the global era. That is an evolutionary lag error of epic proportions—but one that is eminently addressable.

With inclusive stakeholding as part of a larger roadmap to help us find our way home, we have the opportunity to reimagine our social, political, and economic institutions as aligned interests and congruent goals.[242] If we succeed, our transformative journey from the kin village to the global village will be complete. It turns out that our hero's journey was not to pursue our own journey but to steward each other's journeys homeward. We will return not as heroes or as the prodigal sons and daughters, but as both. The two sets of footprints we leave in the wilderness will be one set, not because one is carrying the other but because we're headed home together again.

242. See https://www.prnewswire.com/news-releases/yun-family-foundation-announces-million-dollar-prize-for-social-innovations-300811855.html.

Revolutionary Thinking

Before the time of written words, oral stories were like people: older versions died, and evolved versions emerged that suited changing contexts.

All that changed with the invention of writing.

Written stories could outlive the storytellers and compete with their descendants' stories. The Gutenberg press accelerated the replication of written stories and enabled the masses to access a growing diversity of stories and to formulate their own views.

It was in this setting that a particular book collector in Kraków used the power of books to challenge one of the great ancient stories: that we (planet earth) are the center of the universe. Nicolaus Copernicus amassed a sizable library of astronomy books during the late 15th century, and in 1549 he published *On the Revolutions of Heavenly Spheres,* his own synthesis of newly unleashed information. In his book, Copernicus transformed a convoluted, geocentric model of planetary motion into the elegant heliocentric model that we have today. All prior stories had us as the center of existence, thus the notion that our lives were orbiting around a body other than our own planet was nothing short of earth shattering.

Indeed, the conceptual reframing of existing, seemingly esoteric data to explain the revolution of celestial bodies was so radical that the word "revolution" became synonymous with the now-familiar notion of overthrowing an established system. Dominoes have been falling ever since . . . think the American Revolution, French Revolution, the Internet Revolution, the Blockchain Revolution, and the Artificial Intelligence Revolution.

Today, we are a quarter century into the Internet Revolution. As profound as the impact of Gutenberg's printing press was on the spread of ideas that helped spark the Eurocentric Renaissance, it pales in comparison to the liberation of knowledge achieved with the Internet. This begs the question, "Who will be the Copernicus of our time? Our da Vinci? Our Michelangelo?"

On the one hand, looking to the Copernican Revolution for inspiration may feel like a stale analogy we've moved far beyond; our understanding of astronomy and physics is—with due apology—light years ahead of where we were in the middle of the last millennium.

On the other hand, the Copernican Revolution might just be the perfect analogy for the revolution of our social contract with each other toward eusociality. The social version of the Copernican Revolution could do to egocentrism what the astronomical version did to its anagramic cousin, geocentrism: make someone else the central star of our lives.[243]

243. Today, ecocentrism or geocentrism—putting earth first—has become an aspiration for a civilization struggling with egocentrism.

Interdependence Day

"Freedom is what you do with what's been done to you."

—Jean Paul Satre

Independence Day in the United States is a federal holiday celebrating the colonies' adoption of the Declaration of Independence from the British Crown on July 4, 1776. The American Revolution was sparked in part by the ideal of individual rights espoused by liberal thinkers such as John Locke, Jean-Jacques Rousseau, and Charles-Louis de Secondat, Baron de La Brède et de Montesquieu. The holiday is commonly associated with fireworks and political speeches that commemorate the history, government, and traditions of the United States.

The Declaration of Independence was no doubt a monumental event, but how do we reconcile the celebration of independence with the reality that, a quarter of a millennium later, Britain is now a close ally in the emerging global village? Does it make sense to keep throwing parades about signing divorce papers when we are, in fact, living together again?

Indeed, all nations today are living together in a highly interconnected world. Our fates are intertwined, as individuals and nations, like never before. From ecological impact to interdigitated financial systems, we all have a collective interest in managing risks and opportunities across the planet.

Here are some fundamental questions. Are we going to use this connectivity to scale our localized self-dealing, extractive behaviors to a global level? Are we going to stand by and watch digital versions of imperialism, colonialism, and the Crusades take hold at the expense of the many? Are we going to keep feeding the self-expanding beasts in their race to the middle? Or are we going to use global connectivity to spread the very best of our human values? Whatever we do, the stakes have never been higher.

Our mind's tendency to espouse separatism—the way we did with the separation of good and evil in Eden, the way we did with Beowulf and Grendel, the way we have done in most every story we've told across the millennia—is not only illusory but dangerous.

On the other hand, embracing interdependence may be our most significant existential hope. Imagine global leaders signing a Declaration of Interdependence, perhaps on July 4, 2026—the 250th birthday of America. Imagine a global holiday called *Interdependence Day.*

Psychology 2.0

Holidays are important cultural rituals. They connect us to deeper tribal traditions related to birth, death, harvest, atonement, sacrifice, love, and the passage of time. Who among us remembers that the root meaning of the word "holiday" is holy day? Today, in the West, the hallmark of a holiday is more often than not a material transaction, like a Hallmark card—another example of an ancient virtue commodified by the race to the bottom line. If we are to promote the notion of celebrating our interdependence, it will take more than exchanging cardboard reminders once a year that all but report our negligence the other 364 days.

Is there a way to turn the appreciation of our interdependence into an everyday cultural norm? We believe so absolutely, and we believe we can do so using tools already available to us.

To begin, we envision leveraging the growing public interest in the field of psychology. Once the domain of the elite or the fringe, today even mainstream folks are taking an interest in understanding the habits of the mind and how they can help shape behaviors. But the field of psychology itself is due for an update.

For starters, we need to get out of our own heads.

The emergence of first-person psychology led by Freud and Maslow—the obsession with selfhood and self-actualization—is a symptom of solitarity's triumph over solidarity. The cultural reprogramming of our social contract, then, begins with extending the field of psychology from the first-person perspective to one that also includes second- and third-person perspectives. We call this Psychology 2.0.

Life is an intensely personal and most often self-centered experience. Psychology 2.0 will be about including the interests of someone else in that experience. To give you a sense of our blind spots, the basic four-quadrant chart below shows how Person A reacts to the experience of Person B—another human being:

	Person B is successful	Person B is suffering
If Person A's response to B's state is happiness, then it is called:	?	*Schadenfreude*
If Person A's response to B's state is sadness, then it is called:	*Freudenschade* (envy)	*Compassion*

Schadenfreude and *"Freudenschade"* are psychological expressions of the zero-sum-game mentality we've all experienced.[244] The former is a happy feeling caused by another's misfortune. The latter is a sad feeling caused by another's success.

Compassion is the sadness we feel for the suffering of another person.

What, then, is the term for a person experiencing happiness for the success of another person?

Exactly.

If you were to ask the same question in a room containing one hundred people, it would be a small miracle if even one person called out the word "compersion," which Wikipedia defines as "an empathetic state of happiness and joy experienced when another individual experiences happiness and joy." A word for this feeling is elusive in other cultures as well, and the concept has remained unnamed in many languages. Most Buddhists, for example, are unfamiliar with the word "mudita," which describes the Buddhist concept of vicarious joy.

In other words, one of the four most basic psychological experiences we have in the context of a second person is still awaiting its recognition in many vocabularies. That is a stunning omission.

And yet, compersion is a universal experience in one particular context: parents' experience of their children. Parents experience a quiet (and sometimes not so quiet) joy when their child succeeds. Evolution selected our nature to nurture and encoded this emotional response as a reward reflex to promote

244. See Moffa, M. (2013, December 12). "Which Is Worse: Professional Schadenfreude or 'Freudenschade'?" *Recruiter;* https://www.recruiter.com/i/which-is-worse-professional-schadenfreude-or-freudenschade/; Sivanandam, N. (2006, April 28). "Freudenschade." *The Stanford Daily;* https://web.archive.org/web/20080513182307/http://daily.stanford.edu/article/2006/4/28/freudenschade.

inclusive fitness. High-alignment relationships such as family, however, are less common in today's low-alignment world.

Compersion and compassion are the inherent psychological dynamics of high alignment—a.k.a. kin skin in the game. Is there a way to make these experiences and emotions far more common than Freudenschade and Schadenfreude in today's global society? Compassion is a well-known and widely embraced emotion that is often preached and practiced. Its prevalence is related to the word's frequent appearance in popular culture. Imagine if we could promote the word "compersion" to that level of familiarity and high regard.

That would make us happy.

Compassion and compersion are considered precursor concepts to empathy and are analogous to many notions that appear in all cultures, including the Golden Rule: "Do unto others as you would have them do unto you."[245] The Golden Rule can be thought of as a social codification of inclusive fitness: a mother's kind act toward her child is partially a kind act to herself, given her 50 percent vested interest in the child's genes.

What's lacking from that framing is the experience from the perspective of the recipient, the second- or third-person perspective. No word exists to denote the feeling of *seeking* empathy—a fundamental human experience. Many other fundamental experiences of second-person psychology also have not been named.

Here are some examples. We know a lot about envy, but we don't know much about the psychology of wanting to *be* envied. So much of human pursuit today, including posting on social media, reflects the desire to produce a feeling of envy in others—the feeling of wanting to be popular. Yet we don't have a word for that feeling.

Similarly, no English word precisely captures the traits of seeking compassion, seeking to be understood, seeking validation, or seeking to be the object of

245. See https://en.wikipedia.org/wiki/Golden_Rule.

curiosity. These are definitional phrases awaiting the invention of precise neologisms. The closest approximation of a word that captures the general feelings of seeking empathy, compassion, understanding, curiosity, or validation is "needy," a pejorative term that doesn't do justice to these concepts.

Systems psychology—our Psychology 2.0, or inclusive psychology—remains hugely underdeveloped.[246] As Freud and subsequent psychologists did for their field, we can kickstart inclusive psychology by starting to name the basic second- and third-person psychological phenomena described above.

Assigning neologisms to phenomena has a way of awakening consciousness, thus a cultural shift is possible by creating a lexicon that makes us more conscious and mindful of the experience of others. Creating a new vocabulary for inclusive psychology could promote awareness of others in our lives, just as the creation of terms such as "self-help," "self-advocacy," and "self-love" did to form the consciousness of the self movement.

In some cases, redefining familiar words in new ways can help create a new consciousness. Too many so-called leaders (think "influencers") of today try to amass followers. Imagine instead a world in which the definition of leadership was "to steward the leadership potential of others." That ethos was genetically codified in the kin tribe of yore, but it has significantly faded in today's low-alignment communities and institutions—except in parenting.

The most important aspect of a concept is not the concept itself but how it is used. Machine learning algorithms can be trained to be empathetic to human users—that is, to understand what a person is experiencing—and then use that information to exploit humans instead of serving them. These tools are not inherently the issue; the intentions of the humans behind them are.

In this regard, syntax or grammar can be used in profoundly new ways to

246. See Auerswald, E. H. (1998). "Interdisciplinary versus Ecological Approach." *Families, Systems, & Health,* 16 (3): 299-308; http://psycnet.apa.org/record/1998-12339-008.

promote greater awareness of our responsibility to others instead of focusing on others' responsibility to us (i.e., our entitlements).

Here's an example. Think about world issues and find a way to turn a "they statement" into an "I statement." This can turn a depressing third-person situation into an item of personal responsibility and action. For example, "California is not doing enough to keep the highways clean" can be restated as, "I need to adopt my neighborhood highway so I can keep it clean." The reverse process is cool too. When something good happens, turn an "I statement" into a "you statement": "I scored the game-winning goal" becomes, "Your pass led to the game-winning goal." These processes help us look for people to appreciate and enable us to include others in our story.

Whereas our words have the power to shape our culture, we have the power to create new words and new uses of existing words that can help shift our psychological frame from individuality to interdependence and inclusivity.

When it comes to psychology, it's not just me, it's also you.

Since humanity shifted from eusociality to dissociality, the trend has been to externalize responsibility (thus producing pollution) and internalize love (the self-help movement). Shifting from exclusive stakeholding to inclusive stakeholding is a reversal of this trend: to internalize responsibility (the current externalities) and externalize more love.

Respect shifts from something one gets to something one gives. There's more focus on the rights of others, in addition to our own. There's less blaming and more self-accountability. Indeed, it is possible to espouse the mindset of pre-forgiving everyone despite the inevitable transgressions. Organization charts would flip upside down such that the stewards (or the genuine leaders) sit at the bottom, with each layer of the organization bearing weight on behalf of the team instead of standing on the team's shoulders. Instead of asking for golden parachutes that hold companies hostage, we'd volunteer to wear lead parachutes to leave the rewards behind for others.

The focus of relationships shifts from the question, "How can you help me be successful?" to "How can I help you be successful?"

In his inaugural address, President John F. Kennedy beseeched his countrymen, "Ask not what your country can do for you; ask what you can do for your country." Such antimetaboles can be used as a tool to reprogram the public to become more aware of a contrarian notion. It was thus, in the throes of the Cold War and another conflict that was heating up, that Kennedy launched the Peace Corps.

We might have preferred a version of that statement that better reflects the competing dualities of a memetic parallax—that we are the hero and the beast—by inserting the words "only" and "also": "Ask not *only* what your country can do for you; ask *also* what you can do for your country."

But that's being a bit nitpicky, like pointing out that the heliocentric model is also off target, given the reality that the sun and the earth revolve around each other's center of gravity.[247] Or like pointing out that the founder of the Peace Corps also took the world to the brink of a nuclear war.[248]

For the moment, anyone who can point the world to a contrarian truth, as Kennedy did with his antimetabole, is at least helping split a consensus mob into a memetic parallax so that our responsibility to others becomes more apparent. That alone is one small step for mankind. It would be a giant leap to recognize duality as the enduring truth.

247. And in relationship to the center of gravity of all other heavenly bodies.
248. See https://en.wikipedia.org/wiki/Cuban_Missile_Crisis.

I Know Why the Songbirds Sing

Why do we love familiar songs? Why do they lift our spirits?

Composer Leonard Meyer said familiar music feels good because it satisfies an expected pattern. But taking that proximate reason to explain the ultimate purpose of why we love familiar songs is—to riff off Mick Jagger's familiar refrain—not entirely satisfying.

Let's do another take. With apologies to Keats, the sweetest unheard truth about familiar melodies is that they evoke the same neural reflex as a baby hearing Mom's voice.

And why are babies' ears tuned to Mom's voice? The truth is that fake news is as old as life on earth, and who said it is more relevant than what was said. As the saying goes, "Life doesn't come with a manual. It comes with a Mom." Evolution wired us to trust Mom's voice, and the voices of kin, because they have the most kin skin in the game.

Why is that important? Let's rewind a bit. Songs have been around a lot longer than words. Imagine you are a songbird lost in the woods. Everyone's a stranger, trying to sell you something—or to borrow further from the Stones—you're riding around the world and everyone's telling you useless information that's supposed to fire your imagination.

Then, one day, you hear a familiar song from long ago. The word "familiar" comes from "family." Everyone's song is a unique acoustic fingerprint, and hearing a familiar melody means you are almost home again. Imagine how good that feels.

But nature never foresaw an evolutionary future of recordable music, where early exposure can hook a fan for a lifetime. The risk is that industrialists could decide to use that affection to peddle whatever lyrics or products sell best, even if it means spreading hate, materialism, and selfishness.

But artists could also use music as an instrument of change. Years ago, my father, Sung Hee Yun, one of the founders of the Yun Family Foundation, ended up on Walter Cronkite's TV show because he put education into pop songs—work that later led him to collaborate with entertainers around the world. Yun had this to say about the power of music: People like to sing out loud when they hear familiar songs, which reinforces messages. People like to sing together, which spreads messages.

Most importantly, messages packaged in music can bypass mental defenses and reach the heart. Even the lyrics of "Imagine" needed the vessel of music to bring the words home.

In many ways, Yun was a reincarnation of the troubadours from a thousand years ago who led a rebellion using music to resurface the ideals of a forgotten time. So was Keats, and so were the great artists of the revolutionary 1960s.

Now that we have an answer, my friends, as to why we blow songs into the wind, it begs the next question: who will be the troubadours of today who leave a playlist for the ages that will stream across the airwaves throughout time?

Oh My, My Captain

Stewardship is defined as the responsibility to shepherd and safeguard the interests of others. Many people are in positions of stewardship in public corporations, private enterprises, and charitable organizations. Warren Buffett, perhaps the most widely followed business leader of our time, speaks often of the importance of stewardship in business. It is striking, then, to see the dearth of courses discussing the concept of stewardship at the top U.S. business schools. In the 2013-2014 online course catalogs of the top five MBA programs (as ranked by *U.S. News & World Report*), "stewardship" is nowhere to be found in any course title.[249] To put this into context, each of these schools offered at least five classes with the word "leadership" in the title. In the detailed descriptions of courses offered at the number-one ranked business school in the same report, the word "leadership" appears 108 times. The word "stewardship" is not mentioned once. The closest mention of the word appears in the context of how to steward *yourself,* in a course titled "Leading Your Life."

Fortunately, stewardship—the concept, inherited from the kin tribe age, that preceded the concept of leadership—is an innate core trait. It is observed throughout nature, as exemplified by the nurturing instincts of parents and service behaviors among kin. Such behaviors were evolutionarily selected to meet the needs of social species, such as humans, that grouped around closely related members, or at least did so during the prehistoric age that shaped our selection. It is our nature to nurture, and in the Darwinian calculus, socially altruistic behaviors toward one's kin can enhance one's own evolutionary inclusive fitness.

While such behaviors were likely selected and proved beneficial when humans lived in kin-based tribes, in the modern world of easy mobility and accelerating social dispersal, communities typically develop around diverse, non-kin populations. Many of our factory-setting behaviors, including the degree to which

249. See "Best Business Schools." (2019). *U.S. News & World Report;* https://www.usnews.com/best-graduate-schools/top-business-schools/mba-rankings.

we trust our fiduciaries and agents, are rendered maladaptive in a world of high relationship liquidity.

Given this evolutionary dislocation, stewardship of non-kin is now a moral choice rather than a naturally adaptive behavior that fits Hamilton's rule. These days, we all too often feel forsaken by those we call leaders. A mere century ago, Captain Edward Smith went down with his doomed ship, the RMS *Titanic*. Facing similar situations, the current-day captains of *MV Sewol* and *Costa Concordia* abandoned their ships, leaving their passengers to a disastrous fate. Although these are isolated incidents, they provoke a sinking feeling about the broader fate of civil society.

As a species, we've collectively set sail toward becoming a diverse, mobile, and global community, and there is no turning back. To meet the social challenges that accompany this new era, it is more vital and urgent than ever to start a meaningful conversation about stewardship and inclusive stakeholding. We hope our world's stewards can *lead* the way in starting this conversation.

Internalizing the Externalities

If exclusive stakeholding promotes dissociality, can inclusive stakeholding promote eusociality? And how might the latter model begin to be implemented?

To start, one can reimagine relationships based on transactions as those based on investment.

Here is an example. Most private health insurers currently have a one-year, Tinder-type "trading" relationship with their clients. If the insurer spends dollars on a client's long-term health, such as on preventive care, the long-term economic benefits of future health savings do not accrue to the insurer. This lack of inclusive stakeholding misaligns the incentives between the insurer and the client.

Imagine that the private insurer was instead paid in the following way: a small administrative fee up front, plus a 5 percent annual earnout based on the client's annual health savings (actuarial predictions minus actual spend on healthcare) over each of the ensuing ten years. This payout could be made through a futures contract or on the blockchain through a smart contract. The insurer would thus become an investor in their client's health, where the vested stakeholding would more effectively align incentives between the client and the insurer.

Imagine teachers being similarly rewarded—for example, with tokens of appreciation on the blockchain—for contributions their former pupils make to the world later in their lives. The amounts would be small so that teachers would still have the incentive to teach to the whole class rather than focusing on the next "thoroughbred." This not only would enable teachers to be motivated stakeholders in bettering the future but would reward them for helping the world. These rewards could extend to their being recognized when their students achieve public recognition: imagine Nobel laureate ceremonies featuring the names of all the honorees' teachers.

The more general case is for giving providers a long-term vested interest in the success of their customers.

That model can also be turned two ways. Customers of a company can be given a long-term vested interest in the success of the providers; to some extent, loyalty programs lean in this direction. But reducing, not increasing, the variety of tiers of co-ownership (preferred shares versus common shares in capitalization) would reduce conflict of interest among stakeholders. The idea of customers receiving an actual stake (in the form of stock) in a company with every transactional purchase was debated in the 1990s, but it conflicted with SEC rules. That doesn't mean the rules couldn't be changed. Cryptocurrency utility tokens are the latest incarnation of this idea.

But a core issue of these limited implementations of inclusive stakeholding is the existence of excluded stakeholders—perhaps more distant and less obvious relative to the orbit of the company—who pay a larger and larger price. The issue is not dissimilar to the following paradox in group theory: increasing the number of kids invited to a birthday party can increase the degree of exclusion felt by the uninvited.

This is especially the case when the excluded stakeholder doesn't have a voice, such as the environment, animals, plants, and future unborn citizens. In a world built on exclusive stakeholding, the free market naturally drives itself to profit off these voiceless stakeholders while saddling them with the costs: pollution, animal rights abuses, unsustainable agriculture, ponzi scheme pension funds, the federal deficit, depletion of the commons, etc.

In some cases, the potentially excluded stakeholder of a company is obvious. Abuse of those excluded stakeholders represents an externality to the company, and granting those stakeholders stock in the company is a way to internalize the externality. The abused grow in power proportionally to the growth of the abuser's power; this contrasts with how the world is set up today—toward runaway power polarization.

But in many cases, properly accounting for the excluded stakeholder is hard if not impossible. Even accounting for the good intentions of humans, trying to account properly for all the actual effects current corporate actions have on potentially excluded stakeholders of the future may seem like a fool's errand.

One way to escape this futile loop is to acknowledge the imperfections of such a model, make the best effort today, and hope that future denizens can continually course-correct the world's trajectory.

Another potential solution is for companies to allocate 5 percent stock in the company to a blind trust to deal with the unforeseen and unintended harm caused downstream to future stakeholders. If the corporation is successful and grows in value, then the value of the stock owned by the blind trust grows proportionally, giving the latter the economic power to clean up the untoward effects of the company's success. The governance of a blind trust formed from the bones of the original company would be independent of the governance of the original company.

Moreover, the blind trust would have the same shareholder rights as every other stakeholder of the original company, including voting rights. Thus, instead of merely having the downstream economic power to clean up, say, the adverse effects on environmental stakeholders caused by the original company, the board of the blind trust can speak on behalf of those blinded stakeholders *before* harmful actions are taken.

Including the voices of potentially excluded stakeholders could help steward companies through turbulent waters and toward long-term success. One could, therefore, make the argument that inclusive stakeholding should be embraced by corporations because it is good business. By contrast, many companies—Big Banks, Big Food, Big Tech, etc.—have learned the hard way that exclusive stakeholding can create a myriad of downstream problems for the business.

Finally, the public treasury could be internalized into the capitalization table of private corporations.

On the other hand, some see private profits as corporations' unfair extraction from the commons and view taxes as a way to redistribute value to the commons. Each side is often blinded to the merits of the other side's arguments while failing to see the weakness of its own. The reality is that self-interest is the common enemy that makes all public and private institutions fall well short of serving the people effectively.

Under the inclusive stakeholding model, the potential corporate tax owed to federal and state treasuries could be reimagined as equity stakes granted from private corporate treasuries to the public treasuries. While the distinction may appear nominal, these different modalities for rewarding the public for providing the commons to private corporations create completely different feelings, attitudes, and perceptions. Paying taxes may feel like an extraction by a counterparty, whereas granting stock to public treasuries may feel akin to forming a partnership. The alignment of interests between the private treasuries and public treasuries promotes a feeling that each side is rooting for the success of the other.[250]

As it stands today, public treasuries have collected very little in taxes from Amazon during its entire existence. Meanwhile, shareholders have been rewarded by massive share price appreciation. Such conflicts of interest between private treasuries and public treasuries could be mitigated in the inclusive stakeholding model. Imagine if state and federal treasuries owned stock in corporations as a philosophical alternative to taxes: how would mutual stakeholding between public and private treasuries change the relationship between the public and private sectors? What externalities would such arrangements mitigate and which ones would they create?

Currently, taxes are seen by some as an unfair extraction from corporations by misguided public servants. Many don't trust that providing more funding to the government to address issues or allowing state ownership of private enterprises are the best ideas. In our view, the fundamental weakness of current governance models is the lack of inclusive stakeholding among public sector actors. Since perverse incentives are pervasive, public sector inclusive stakeholding innovations would have to accompany private sector ones to better align interest across the entire system.

There are many potential implementations of this model to be considered, and the goal here is not to be proscriptive—except to guard against undue exercise of

250. Part of the aim here is to promote the emotional transformation that occurs when former counterparties are realigned as partners. Just as labor is currently treated as an expense on corporate financial statements, so too is tax.

ownership control by the state. Our argument is more of a philosophical nature than economic. As excluded stakeholders become included, externalities become internalized. Former counterparties become partners the way former warring houses become allies when their children marry.

During the Industrial Revolution, labor and capital were mutually excluded stakeholders that waged battle over wages and other issues. Silicon Valley shareholders internalized labor and turned their workers into partners through stock options. That revolution was so bloodless that history has largely failed to even mention this revolutionary innovation as the seminal contribution of Silicon Valley.

The productive potential of extending this inclusive model of stakeholding to the commons—imagine corporations turning the conflicts of interest between their private treasuries and public treasuries into an alignment of interests—could be similarly vast. Then imagine such corporations competing with each other in a free market race to the top instead of a race to the bottom.

Universal Basic Stakeholding

"The future is here. It's just not widely distributed yet."

— William Gibson

Universal basic stakeholding (UBS) is a generalized theory of granting all individuals skin in the game in a collective society. The stake granted to the individual stakeholders of a system is called a universal basic stake. Alignment of interests, congruent goals, and interdependent inclusiveness are the fundamental values of UBS.

UBS differs from universal basic income in that UBS is a vested interest *in the system* and not a payment taken *out of the system*. An argument can be made that universal basic income is indirectly tied to the success of a nation, the way an employee's salary is indirectly tied to the fate of a company. However, direct ties that people can appreciate, by virtue of a stake, is the essential basis of appreciation of value.

The model is similar to stock grants made to the employees of a corporation, except that all stakeholders of a company, not just employees, would have a stake. One example of the implementation of UBS is a tradable blockchain token granted to every citizen of a nation or of the world. Imagine multiple layers of stakes, such as stakes held by and between countries to form alliances and to create mutually vested interests to reduce the risk of conflict.

In the broader context, the idea behind UBS is to provide an updated, more robust social algorithm to mimic and replace the bioalgorithm of inclusive fitness that served as the social operating system during the kin-tribe era of human evolution. For most of evolution, the kin tribe was the operating unit of human social systems. Everyone had kin skin in the game in each other's welfare, which kept extractive behaviors in check. In the post-diaspora melting pots of genetic strangers, however, this hardwired programming of inclusive fitness became maladaptive, as people began taking from, rather than investing in, their community of strangers in order to serve their own kin preferentially. Our factory-setting instinct to feed one's family became the basis of mutual extraction among kin networks that co-existed in larger melting pots. That dynamic translated into the emergence of extractive social, political, and economic institutions everywhere in human society.

Frank Herbert, author of *Dune,* famously claimed that feudalism is a natural condition of human beings.[251] He said, "We tend to fall into it, given any chance at all, given the proper stresses or given the proper lack of stresses, and I think we could extract away from that notion that we have many more feudal or tribal aspects in our society than we might've otherwise thought about it."

This view is inconsistent with evolutionary biology. The natural condition of human beings is to live in tribal social groups governed by kin skin in the game. Competition selects for power. Whereas the empowered in a kin tribe are genetically motivated to rule in the best interests of all in the tribe, the empowered in melting pots are genetically motivated to self-deal against the interests of the people. The empowered could represent a political, social, or economic elite that extracts value from the total system, thereby widening wealth inequality.

Today, the factory-setting bioalgorithm of inclusive fitness has been rendered maladaptive in the post-diaspora global village, spilling a myriad of seemingly intractable social issues from Pandora's Jar and posing an existential risk to the future of human evolution. The current system naturally promotes inequality, buffered by payouts to the disadvantaged to attenuate revolutionary ardor. Universal basic income is an example of such a payout. By contrast, UBS, in the form of equity in the collective prosperity, offers a heretofore unprecedented social algorithm for equality: the leading edge pulling up the trailing edge, rather than leaving people behind.

251. See https://www.jstor.org/stable/4240179?seq=1.

Interdependent Capitalism

Editor's Note: Prosociality can be nurtured through the evolution of attitudes. That said, Joseph Henrich noted that prosocial institutions over time collapse at the hands of self-interest. Could harnessing self-interest itself fuel a more enduring form of prosociality?

"Man is born free, and everywhere he is in chains." That's the opening sentence of Jean-Jacques Rousseau's *The Social Contract.* Written in 1762, Rousseau defines man's natural state as being free and happy, and man's loss of autonomy in the modern world as being the root of alienation, injustice, unhappiness, abuse, and war.

We make a different claim.

We claim that man's natural state is interdependence and that, without sufficient mutual vested interest to reflect that interdependence, man's pursuit of autonomy and enlightened self-interest has also been a source of alienation, injustice, unhappiness, abuse, and war.

Is there a way to reconcile these seemingly conflicting philosophies? Absolutely.

The common ground is this. Independence without mutual vested interest can promote the extractive behaviors and abuses we fear. Interdependence without mutual vested interest can also promote the kinds of extractive behaviors and abuses Rousseau feared. Thus, the debate ought not to be about the merits of independence versus interdependence. It ought to be about the lack of mutual vested interest.

Indeed, if mutually vested interest is properly designed as the social contract in an interdependent network, man's innate tendency to pursue self-interest can be harnessed to work for mutual benefit rather than extraction. That is, social algorithms for inclusive stakeholding can be designed to mimic the bioalgorithms of inclusive fitness.[252] The congruence of goals increases with the degree of mutually vested interest.[253]

Can we tie the success of teachers to that of their pupils? Can we tie the success of health insurance, food, and media companies—indeed, of nearly every industry—to that of their customers as well as the public interest? Can we tie the success of elected officials to that of their constituents? Can we find ways to include the environment, future generations, and others who have no voice as stakeholders? Can we redesign physical and online communities around the right mix of diversity and common interests the way kin tribes once did?

We believe that the recent emergence of blockchain and related technologies represents an unprecedented platform to instantiate this vision: to create vested interest among interdependent stakeholders, inclusively defined, the way inclusive fitness did for kin tribes.[254] It would be akin to taking the notion of employee stock options common at Silicon Valley companies and extending it to users, consumers, and other stakeholders.[255]

Establishing congruent goals through stakeholders' vested interests would fundamentally transform every dimension of society.

We hope that, by shedding light on the origins of our modern self-dealing behavior, readers can take a more enlightened position about the way we treat each other. It is our nature to nurture those in whom we have a vested stake. But we can do

252. Inclusive stakeholding is the notion that all potential stakeholders of an institution are given appropriate amounts of skin in the game.
253. For example, if a social media company, advertisers, and users co-owned stakes in each other's success, the incentives for extractive behavior would be proportionally reduced. They would have a higher degree of goal congruence than if they didn't have such mutually vested interests.
254. To date, however, some blockchain initiatives have been characterized by the same self-dealing behaviors of traditional institutions. In fact, generally speaking, applying the powers of technologies to the same human foibles only amplifies the abuses. Other recent examples of this trend include social media companies, artificial intelligence, and robots. On the other hand, these risks could be turned into assets if we created a model of vested interest in the success of others. Rather than developing policies to regulate these runaway institutions, we can design "policies" that promote desired social behaviors within technology protocols (e.g., smart contracts).
255. Blockchain and token economic models are able to supersede the traditional theory of the firm and blur the lines between shareholder, consumer, and other stakeholders in the broader community.

better. We don't have to rely on nature or our upbringing (nurture) to govern our decisions and actions. We can act with personal agency. We can choose to do the right thing for others, simply as a matter of personal choice. Reprogramming culture is part of that.

However, we may not be able to count on personal agency to override people's tendency to pursue self-interest; we believe the key will be to create new social contracts that give people vested interest in the success of others.

To be clear, however, there is no such thing as a perfect system of incentives. Even kin tribes undergo a rapid dilution of alignment as lineage arborizes. It also is perfectly common for even close families to become estranged and scatter. Either way, every house becomes divided over the generations, dispersing into self-replicating units of new tribes. Imprinting on siblings promotes dispersion, outbreeding, and diversity among progeny while deleterious homozygous diseases punish inbreeding. Dispersion, in short, is a feature of social evolution, not a bug. The rapid dilution of shared genetic interest foments greater degrees of competition between groups as the generations unfold, but within each new group, collaboration is the rule.

It seems that some balance of competing interests and aligned interests within social groups would maximize evolutionary fitness across spatial and temporal fractals. In other words, the diversity of competing interests, like the alignment of interests, is a feature—and not necessarily a bug—of complex social systems. Speciation, population diasporas, and blockchain forks result from, and contribute to, the emergence of complex ecosystems featuring a mix of diversity, competition, and collaboration.

There is also no such thing as the "right" mix of these features for any particular ecosystem, as a system will adapt as necessary over time. However, that adaptation won't happen soon enough for humans living today. Furthermore, ignoring the cancer of perverse incentives will only lead to a metastasis of self-dealing behaviors.[256]

256. Perhaps inserting vested interest into the genetic code of rogue cancer cells could help the "beast" domesticate itself back to being a constructive member of the interdependent system called the body.

We need solutions today. Rather than being prescriptive, however, we believe that, in the proper context, solutions can be produced from the collective power of the community by the very elements that power evolution: competition and coopetition. For example, incentive prizes could be designed for nurturing a new model of interdependent capitalism. Competition is a natural force of social systems, and we can leverage this force to nurture the best ideas for social innovation on incentives. We believe in the bottoms-up philosophy that each community—*when properly incentivized to do so*—is in the best position to create incentive structures that serve its greatest good.

Systems Philanthropy

The world's current model for addressing problems can be described as an ecosystem of competing and collaborating organizations with focused objectives: ending hunger, reducing homelessness, supporting the arts, etc. Features of this model include specific objectives, multilateral points of view, and survival-of-the-fittest style interorganizational competition for resources. The specificity of mission statements is critical to giving an effort the focus it needs to achieve the mission's outcome.

However, there is a risk to this model. Given that our planet is an interconnected system of systems, without a systems perspective, solving a "local" issue may, in a whack-a-mole fashion, create a new issue over time. The tendency toward focused missions can blind one to the existence of excluded stakeholders who pay the price of the unintended consequences of an organization's specific efforts. One example is how the efforts to alleviate world hunger in one generation contributed to the global rise of obesity in a subsequent generation.

A key feature of systems philanthropy, then, is inclusive stakeholding, whereby external stakeholders are internalized and accounted for in every mission. One implementation of this philosophy is a management practice whereby the intended and unintended consequences of management actions on all stakeholders, inclusively defined, are diligently considered up front. Thereafter, management decisions can be directed at benefitting all stakeholders. A particularly important group of potentially excluded stakeholders is those without a voice, including the environment, other forms of life, and future generations.

Take the example of future generations. It is essential to factor these future stakeholders into current decisions, lest they lead to intertemporal imperialism (i.e., borrowing from the future to pay the present). The public pension fund models of today fall into this category. Thus, four-dimensional thinking involves not only canvassing a wide spectrum of current stakeholders but considering them longitudinally and dynamically over time.

The best way to mitigate kicking the can down the road, of course, is to identify the root cause of problems and then implement curative Occam's razor solutions, rather than symptomatic treatments. However, that's often a very tall order. Moreover, even when something is "cured," the effect of the cure is still likely to have adverse consequences for the overall system. Such is the inherent nature of all human endeavors.

Another possible implementation of the inclusive stakeholding philosophy is to require that 5 percent of each organization's net profits go into an independently managed trust fund that will be deployed later to address the issues that arise as a consequence of the original organization's efforts. This would help establish intertemporal interdependence among stakeholders over time.

An ideal system would not only thoughtfully address issues on behalf of all stakeholders over time but increase system resilience, much the way a current account surplus builds buffers in a financial system. How do you improve a system's resilience? As with the health of individuals, a system's homeostatic capacity can be enhanced through capacity-building.

The World Economic Forum's current mission statement is, "Improving the state of the world." What if our collective commitment were updated to be, "Improve the state and the resilience of the world"?

Can Eusociality of Our Cells Improve Our Health?

The inclusive stakeholding framework can help explain the emergence of cancerous behaviors between humans, and it also may help us better understand cancer itself.[257]

Based on Hamilton's rule, the same individual can display nurturing behaviors toward kin and extractive behaviors toward non-kin. For example, a tyrant may extract from his people but lovingly nurture his own children. The person hasn't changed between these two scenarios. What has changed is the context; specifically, the tyrant's degree of vested interest in the relationships, which affects congruence of goals, alignment of interests, and degree of interdependence.

The human body can be considered a society of cells with identical genes. The trillions of cells in a normal body run their algorithms as a coordinated, interdependent superorganism. Cells in the body are known to nurture and behave altruistically toward one another: apoptosis is one example, as is the specialization of cells in a multicellular organism.[258] When the network of interconnected algorithms manifests a mutually vested interest in each other's success, the organism thrives.

Intercellular interdependence, however, appears to decline as we get older, and this may contribute to the aging process.[259] Which processes lead to the loss of intercellular interdependence remains to be determined. Perhaps the loss of a central biological clock after pineal involution promotes dyssynchrony among

257. Yun, J., Yun, J., & Yun, C. (2019, February 15). *Interdependent Capitalism: Redesigning the Social Contract through Inclusive Stakeholding* (Kindle edition); https://www.amazon.com/Interdependent-Capitalism-Redesigning-Inclusive-Stakeholding-ebook/dp/B07NSW9SYZ.
258. Williams, S. (2012, June 11). "Researcher Identifies Unusual 'Altruistic' Stem Cell Behavior with Possible Link to Cancer." *Stanford Medicine;* http://med.stanford.edu/news/all-news/2012/06/researcher-identifies-unusual-altruistic-stem-cell-behavior-with-possible-link-to-cancer.html; (2006, September 18). "Why Evolution Drives Some Cells to Altruism." *Phys.org;* https://phys.org/news/2006-09-evolution-cells-altruism.html.
259. Suma, D., Acun, A., Zorlutuna, P., & Vural, D. C. (2018, February 21). *Interdependence Theory of Tissue Failure: Bulk and Boundary Effects.* London: Royal Society Publishing; https://royalsocietypublishing.org/doi/full/10.1098/rsos.171395.

cells. A loss of power, control systems, or repair mechanisms could also contribute to the loss of intercellular interdependence.

Perhaps the loss of interdependence promotes selfish behavior among cells, whereby existing bioalgorithms are subverted to promote extractions. Internalities are externalized, leading to perverse incentives, and competition for resources emerge. Not unlike self-dealing nations, tissues in the body become tribal and extract from the total system. For example, tissues appear to hoard high-density stores of energy, such as fat. Indeed, an increasing degree of adiposity is observed in every tissue of the body as we get older.

Consider cancer through this same lens. When normal cells lose intercellular interdependence, they act more independently. Once independence is established, a conflict of interest develops.

Thereafter, evolution can select for more extractive behaviors. Whereas alignment of interests and competition is a race to the top, malalignment of interests and competition is a race to the bottom. Over many divisions of the cancer cell, selection favors the emergence of ever-more extractive algorithms that drive the cell to become a self-expanding beast, at the expense of the host. The fractal analogies to extractive social, political, and economic institutions are self-evident.

This lens offers a new target for treating cancer: to reestablish interdependence among cells.

One of the conventional theories about how to treat cancer is to attack extractive algorithms in mutant cells. Blocking growth-factor pathways is an example. This approach, however, may prove futile if intercellular interdependence is not reestablished. Even worse, the treatment can backfire by acting as an evolutionary selector for more aggressive algorithms, the way antibiotics can select for resistance.

We hypothesize that reinserting algorithms of intercellular interdependence may allow all those extractive behaviors to self-domesticate into regulated normal behaviors that serve the host rather than the cancer. Essentially, from the perspective of cancer, the host becomes an externality that cancer pollutes and exploits.

We are early in our thinking about this approach. We wonder about injecting oxytocin into tumors to increase a nurturing tendency among cells, since the hormone is known to promote nurturing behaviors in humans. There is some preliminary data on this idea.[260] For these reasons, we believe that the biological basis of intercellular interdependence—and the mechanism of its loss in cancer and aging—warrant further investigation.

Even larger implications come to mind. In the arc of evolutionary history, we were able to become humans after aggregates of selfish bacteria were able to form nearly perfect eusocial systems known as multicellular organisms. Thereafter, we became the superorganism relative to the cells in our body—the hive relative to the subordinate beings.

Over the course of the life of a human body, cellular lineages divide, and, as they do, the cellular lineages arborize to some extent. What ensues bears deeper consideration. As multicellular organisms, including humans, age, our cellular components appear to behave more selfishly. High-density hydrocarbons are extracted from the total system and stored by cells locally in the form of fat. Some cells go even more rogue and exhibit full-on sociopathic behaviors, such as fooling neighboring cells while stealing resources from them. In essence, our body appears to devolve from a eusocial hive of individual cells to a dissocial hive. That's one way to think of an aging body.

A potential fractal recursion emerges from this thinking.

If human aging reflects the transformation of a eusocial hive of cells into a dissocial hive, could restoration of eusociality among the cells in our body reverse the aging process and restore youth?

Similarly, can we think about the gradual emergence of sociopathic social, political, and economic institutions as the natural consequences of an aging civilization superorganism? As replication among lineages proceeds over generations of division, dissociality replaces eusociality as the social algorithm of humanity.

260. Ji, H., Liu, N., Yin, Y., Wang, S. et al. (2018). "Oxytocin Inhibits Ovarian Cancer Metastasis by Repressing the Expression of MMP-2 and VEGF." *Journal of Cancer,* 9 (8): 1379-1384; http://www.jcancer.org/v09p1379.htm.

To extend the fractal metaphor, we can think of the effort to build universal eusociality as an effort to return a flailing, aging civilization to youthful health. There are many lessons to be learned from comparing the efforts of universal eusociality with efforts to reverse human aging.

Self-Driving Revolutions

Can self-driving cars teach us something about self-driving revolutions?

Autonomous vehicles use algorithms to drive themselves safely and efficiently to a desired destination. What if one could type in a desired destination for a social movement and let it drive itself?

On the one hand, that might sound enticing. On the other hand, as with every technology, the fundamental question is not "What can technology do?" but "Whom does it serve?" What if nefarious forces could one day socially engineer individual consciousness and collective culture as easily as it would be to type in a destination on a self-driving car's dashboard? It all comes down to the intentions of the one typing in the destination. Leaving that to fate alone doesn't sound like such a good thing.

But what if that is already happening today?

Indeed, we know it is already happening to some extent. This isn't just a reference to internal and external influences on politics. Think of consumer-product campaigns. Marketing is social engineering intended to shape individual consciousness and collective culture. Software now helps brand managers listen in on the public's response to their product and tune their campaigns accordingly. Not coincidentally, such software is often referred to as a dashboard.

Let's not kid ourselves. Marketing, as social engineering campaigns, is everywhere. It's actually a rat race of social engineering experiments run by institutions competing for your buy-in, and you are the experimental lab rat with your paw on the lever.

But, before you go tearing up the cage, know that this game has been going on since the beginning of humankind. Kin skin in the game is a social engineering program—selected and embedded biologically by evolutionary forces in a decentralized fashion across the hive—to coordinate more cohesive behaviors. Imagine the tables being turned, with larger forces looking in on us the way we look at ant farms. Are we high-agency altruists, or are we just drone pawns running the altruism code?

Seen from that perch, like Alice in *Through the Looking Glass,* we've always been pawns in a larger game. Actually, each of us is a pawn being played by an infinite set of games that are competing among themselves for your power. Social gratifications such as pair bonding, mating, and brood nurturing may make poets swoon, but these are merely proxy rewards for pushing the dopamine lever that keeps the game play of evolution moving forward.

And you are doing it too. Being a pawn in the larger game means your everyday good intentions drive the gears that may very well extract power from the lives of others.

Moreover, you are not just doing it to your peers. You are also a superorganism commanding the individual cells in your body that live and die every day, in your name. Unlike your pawn-like contributions to The Man, at least the cells in the body are aligned with you. Yet these individual cells are not aligned with the cells that make up your neighbor. Again, the same fractal combination of in-group loyalty and outgroup competition is evident. This is the brutal multilevel selection algorithm that has made the world go around in a traffic jam of autonomous living systems.

All that said, we have a choice on multiple levels. We are sentient beings. We don't have to play the game. We may even be able to change the game. We know that we also could just let the game play out as is. Surely eventual mutual destruction is not assured? If we left human nature to run its natural course, would we keep pulling rabbits from hats when pushed to the limit?

But there is an even broader question: can we afford this game of roulette? The risks of changing the game are high, but the risks of not changing it may be far higher. And if we come to the latter conclusion, are we even capable of changing the rules of the game while being its pawns?

This is the entire miracle of the human species.

We as a species are decoding the rules of the game while being trapped inside of it. We are leveraging the natural momentum of the game to change its outcome. We harnessed the Promethean Fire and have harnessed far more since. We are capable of creating technology to re-create ourselves.

But we first must come to terms with the fundamental paradox: individual self-interest doesn't scale as an algorithm of sociality. Our libertarian instincts will continually be surprised at the rise of mob-like superorganisms, composed of individuals' interconnected self-interests, that seize power over individuals like a self-eating snake. Imagine new algorithms for human sociality that scale beyond that of clucking over one's own brood. Imagine algorithms for the self-driving revolution of universal eusociality.

The Ouroboros

Past efforts suggest that trying to get people to understand that we might not be the center of the universe is non-trivial. Self-centrism has an innate appeal that other-centrism can only envy. The battle between them is akin to Goliath versus David, but without a slingshot.

Those who have put their skin in the game to try to get people to accept the idea of putting others at the center of life have paid a large price. Jesus gave up his life. Copernicus feared the potential reaction to his book, and his own reaction upon being presented the first printing was to drop dead.[261] Galileo Galilei, found guilty of heresy for carrying on the Copernican view, was sentenced to indefinite confinement and forced to read seven penitential psalms a week. In the spirit of kin skin in the game, his daughter Maria Celeste relieved him of that punishment by securing ecclesiastical permission to take it on herself.[262]

When defending an underdog position in a memetic parallax, Galileo's fate exemplifies the perils of resistance against self-dealing, self-expanding beasts. The years leading up to his trial were characterized by the typical escalation of mob mentality, unnecessary theater, and polarization of camps. If anything, opposition to heliocentrism became entrenched. It wasn't until after Isaac Newton published *Principia* in 1687—nearly half a century after Galileo's death—that the heliocentric view became generally accepted.[263]

Is there a way for an underdog to win without throwing stones, fighting the resistance, and alienating—or being alienated by—the very opponent they are trying to persuade?

Aikido, the Art of Peace, is a martial arts form known for using the opponent's own force against him.[264] One common maneuver is to use an opponent's

261. Bell, E. T. (1992/1940). *The Development of Mathematics*. New York: Dover.
262. Shea, W. (2006). *The Galileo Affair* [unpublished work]. Grupo de Investigación sobre Ciencia, Razón y Fe.
263. Kobe, D. H. (1998). "Copernicus and Martin Luther: An Encounter between Science and Religion." *American Journal of Physics*, 66, 190.
264. See https://en.wikipedia.org/wiki/Aikido.

momentum to throw him to the mat. Harnessing the adversary's energy enables the actor to fell much larger opponents.

Revolutionizing our social contract will require a fundamental shift in the momentum of our culture. Self-dealing has been the prevailing human behavior for so long, and has been so well rewarded, that its hold on culture is pervasive. Self-expanding beasts are towering over society everywhere on the horizon.

So, what's the big idea here? Slay these self-expanding beasts one by one? That's not a sensible approach, for several reasons. First, it would be futile, as the system would self-select other beasts to rise in their place, like a "whack-a-zombie" apocalypse. Second, the beasts are composed of the exact community of people we are trying to protect. Third, as discussed above, we might in fact be the beast, not the hero. Fourth, we might represent the duality of being both the beast *and* the hero.

A better approach may be first to recognize that the beasts are second-order symptoms, and then to understand that trying to address a second-order symptom before solving the first-order problem will lead to third-order derivative issues.

The first order of business, then, is to create a system of inclusive stakeholding to replace inclusive fitness as the fundamental social contract of humanity.

We are optimistic about this.

We are optimistic because there are a lot of tools at our collective disposal. We are not limited to rewriting fairy tales, like the hero's journey. For a world that has remained a bit flat-footed on innovations that create incentives, there has never been a better environment—and a greater need—in which to offer radical solutions. We are optimistic that the existing forces driving the "self" culture can be redirected toward a self-driving revolution against the culture of self-centeredness—that is, to use everything the beast is good at for the greater good and take advantage of the beast's innate instinct to take advantage.[265]

Even small tilts in incentives can compound through the self-interests of seven

265. The key components of self-driving revolutions are (1) feed the beast its own tail—that is, redirect the natural forces of systems to self-correct the dysfunctions; (2) write the decentralization of power into the movement's self-replicating code—that is, give others the stage, credit, resources, and voice.

billion people and multiple organizational systems layers to produce dramatic prosocial outcomes.

In the ancient tradition of the ouroboros, let the snake eat itself.[266]

266. See https://en.wikipedia.org/wiki/Ouroboros.

Race to the Top

The "tree of life" is a mythological archetype that appears throughout the world's religious and philosophical traditions.[267] A branching tree is also a metaphor for evolutionary speciation, blockchain forks, and many other systems that feature ramifications.

What is a branch point in a tree? Analogous to memetic parallax, a branch point is where ramification—a division—occurs (e.g., the duality of good and evil in the tree of knowledge in Genesis). On a tree, a common central trunk bifurcates into separate limbs, which bifurcate into branches. The process repeats iteratively up to the tips of the leaves.[268] Fractal recursion is evident.

The tree of life can also be a metaphor for the diaspora of a kin tribe over the generations. From a central shared trunk, kin lineage iteratively bifurcates in repeating patterns over time. The further downstream from the central trunk, the more distant the cousins. The processes at each branch point have common features, fractally speaking, just like the tribe units of successive generations.

Conversely, can the diaspora of a kin tribe over the generations also serve as a metaphor for the tree? Is a genomic diaspora observable in the ramification of branches on a single tree the way kin genomics ramify? As might have been predicted, based on this thinking, leaf genome sequences on an individual tree were not found to be identical; they varied systematically as a gradient from the bottom to the top of the tree.[269] Thus, the genetic variance among branches increases along with the degree of ramification. Each tree exhibits microchimerism.[270]

267. See https://en.wikipedia.org/wiki/Tree_of_life.

268. Apoptosis or programmed death was discovered in the abscission of leaves. The leaves become part of the biological compost in the vicinity of its own roots, contributing substrate that is recycled for the renewal of life in the general direction of evolutionary progress. One can think about phenoptosis (programmed death) among humans the same way.

269. See Diwan, D., Komazaki, S., Suzuki, M., Nemoto, N. et al. (2014, February 19). "Systematic Genome Sequence Differences among Leaf Cells within Individual Trees." *BMC Genomics;* https://www.ncbi.nlm.nih.gov/pmc/articles/PMC3937000/.

270. One wonders if the cell turnover within an animal similarly creates microchimerism (which is known)

The tree of life metaphor can help distill other themes explored in this book. Given the interdependence among dividing limbs, the bifurcations are dehybridizations characterized by some degree of competition, overall goal congruence, and vested interest in each other's success. As a result, the Darwinian competition among even the most distal leaves are as benign as Little League Baseball: there is enough competitive dynamic to help nurture the selection of beneficial traits among the distal leafy cousins but not enough competition to seriously hurt each other.[271] The distal branches can be thought of as genetically different cousins who compete with each other for resources provided by the central trunk below; they grow upward and past each other as they compete for sunlight from above.

One can think of competition for resources from the tree trunk as competition for our mother's attention and the vertical reach for sunlight as competition for our father's attention.[272] But all competition among the tree parts is subordinate to their shared fate, their interdependence. The close alignment and competition together promote growth—a literal race to the top.

and in-organism genetic diaspora over the generations of cells. One wonders if the loss of eusociality and the emergence of dissociality through the cell division process contribute to the aging process and the emergence of cancer?

Though rare, microchimerism is also observed in animals, including humans. We are built through iterations of cell divisions among lineages, starting with the embryo. As genomic tests advance, perhaps we will learn that microchimerism is more common in animals than is observable today. Or, we might learn that genomic ramification—the decline of genetic alignment and the loss of vested interest in the success of other cells—is too problematic for animals to maintain as a whole so that genetic variance is policed out of the body by the immune system. Genetic ramification could be a factor in a wide variety of human ailments, including aging, autoimmunity, and cancer. Individual cells in a 60-year-old human could be seen as the tips of tree branches competing rather than collaborating with each other.

A human cancer cell, then, is a division gone rogue. Perhaps the immune system clears most cells that have gone rogue, but it is not always successful. Furthermore, a cancer cell can subvert existing host pathways and hijack them to self-deal at the expense of the host. The cancer becomes a self-expanding beast that feeds on the host—a former community member that grows by feeding off the community. The parallels to institutions in our society today are self-evident.

271. Nature exhibits a diversity of approaches in mixing the degree of competition and alignment in tribes. A tree's ratio of collaboration and competition and that of a human tribe are different, but the two can be seen as variations on the theme of a parallax of approaches in evolutionary programming among the plant and animal kingdoms.

272. If we continue with the analogy, one implication of the attraction to the sun is that the evolutionary progress of life on earth may not be just a self-referential unfolding or emergence but a bending toward a certain destiny we are being attracted or pulled to. More generally, given the fractal nature of nature and the expression of the universe as emerging from recursive functions, we might be able to deduce the totality of the universe by looking at the organization and functions of a tree relative to its surroundings.

There is a lesson here for humanity. Microchimerism is also observed in animals, including humans, but it is rare.[273] The lower tolerance of the genetic diaspora of cell lineage in animals than in trees may relate to the degree of interdependence. If the limbs of a tree are seen as a collection of conjoined cousins, when one cousin dies, the remaining branches not only survive, they thrive.[274] The greater tolerance of genetic variance in trees versus animals is evident in the greater ease of grafting transplants. Whereas human cousin groups can tolerate even greater genomic diversity than a tree—their bodies and their fate are not intertwined—when one-half of conjoined humans dies, the whole typically dies.[275] Thus it seems likely that microchimerism is less frequent in animals than in trees, and less frequent in trees than in a kin tribe as a superorganism.

Within eusocial human tribes, however, the story is different. The tribe can be thought of as a microchimera, where the diversity of imperfect genetic overlap is a feature, not a bug. Yet, we have also seen that, as the mutual kin skin in the game declines, the falling degree of genetic overlap does become a bug. The interdependence within eusocial tribes turns into the independence within dissocial tribes, and competition displaces collaboration. As a consequence, sociopathic behaviors proliferate.

However, we make the case here that we can re-establish humanity's interdependence—the trunk of the tree of life of humanity—by updating the bioalgorithms of inclusive fitness with the social algorithms of inclusive stakeholding. Creating this interdependence is akin to forming a common trunk for all of humanity. Thereafter, human dissociality returns to eusociality.

At scale, we can build networks of interdependent stakeholders to create a social economy, as Facebook did for social media but much, much bigger. Yet

273. See https://www.ncbi.nlm.nih.gov/pubmed/17917028; http://citeseerx.ist.psu.edu/viewdoc/download?doi=10.1.1.149.9001&rep=rep1&type=pdf.

274. The fate of conjoined humans is similar to when a common trunk of two limbs of a tree dies: the whole unit perishes.

275. That said, there may be more contagious signaling across non-conjoined humans than we realize. Are there senescence pheromones among humans, like the coordinated contagion of senescence through ethylene among leaves in a forest? Here's a related question: why did nature select for conscious signaling in some cases (spoken word) versus subconscious signaling in other cases (pheromones)? Perhaps there is adaptive benefit in the subconscious coordination of group behaviors.

rather than being prescriptive about how to redesign our institutions through inclusive stakeholding, we rely on the ouroboros approach: using incentive competitions to nurture inclusive stakeholding innovations. That is to say, with congruent goals, we can use the exact forces conspiring against the interests of the people today to work for the people in the future. Whereas malalignment with competition is a race to the bottom, alignment with competition—like the tree of life—is a race to the top.

Life, Itself

At the beginning of this book, we praised parental love. Then we chafed at the nepotism it spawned. Next we lauded moral traditions homilizing kindness. Then we reproached the abuses committed in their names. We subsequently celebrated competition for shaping prosperity. Then we impugned it for creating a world in which we gorge on high-fructose corn syrup and the Kardashians. You can feel the tension of these self-contradictions. In one minute, an institution is the hero in our story. In the next, it is the beast we are attacking. The distinction between hero and beast comes across indistinctly at best.

No concept embodies the tension inherent in this duality better than the Chinese *taijitu* symbol of yin and yang. Yin and yang beautifully express both symmetry and opposition. The interdependent mutuality of yin and yang is dynamic, with each force begetting the other in a never-ending cycle. According to the philosophy represented by the yin-yang symbol, all aspects of existence flow from this cycle of opposing forces, including flow itself.

The perspective offered in this book, however, suggests that the existence of the yin-yang symbol can itself be seen as the downstream result of something more upstream: the tendency of otherness to emerge from oneness through the recursion of nature. Our dawning awareness of this emergent duality was enough for us to be sent packing from our Edens and from our gohyangs.

Thereafter, the chapters of history wrote themselves. From the point of view of the human condition, harnessing the Promethean Fire was the first-order event; the shift from eusociality to dissociality was a second-order event; the emergence of dissocial superorganisms that ruled over, instead of on behalf of, the people was a third-order event; the stampede among dissocial superorganisms sparking self-consuming revolutions was a fourth-order event; the shift to subverting revolutions and dehumanizing humans into pawns who willingly toil on behalf of the system was a fifth-order event.

To wake up in this kind of world has been bewildering. To one day see power structures through the looking glass as beasts—distorted versions of their purported intent—was disheartening. To finally recognize ourselves as that beast was an even worse funhouse-mirror moment.

Yet, when philosophers and artists consider the possibility that humans are living in a simulation, they tend to evoke the image of an outside agent running a program that we can't imagine. A different way to imagine the simulation is to see every power structure as an inverted version of their original purpose, operating through nothing more than the aggregate sum of our own self-interests. We do not have to evoke a higher presence orchestrating from behind the curtain to explain our surreal experience. The whole story requires no story other than our own.

And that's the good news.

We're at that "tap your heels together three times" moment when we realize that we have had the power to change our story all along. We already possess the tools, the agency, and the historical moment to be the yin to the yang, to leverage the momentum of the system against itself. To use incentive competitions to undistort perverse incentives is to nudge the beast closer to where it already leans. To use stories to undistort the story market—to tell Grendel he is the hero and not the beast—is to feed the snake its own tail, if you will.

The ultimate way for this book to bow out of these larger stories would be to have it eat its own words. Sure, the proximate aim of our journey on these pages has been to render a landscape full of fertile possibilities so that we can forge ahead with lining up the gears of a self-fulfilling revolution toward global eusociality.

Yet there is even better news . . .

Here is a brief history of human kindness.

In its more primal form, human kindness occurs when people have kin skin in the game. After all, kin is the root word of kindness.

With the advent of moral, spiritual, and religious philosophies, another driver of human kindness was birthed—kindness performed for rewards in the existential, spiritual, or afterlife realms.

With the emergence of transactional and economic incentives, kindness for profit emerged as another category of beneficial behavior toward others.

All of these forms of kindness, however, suggest that we are kindness mercenaries—that we perform acts of kindness because they benefit *us*. Moreover, rarely do we slow down enough to realize that the instant an act of kindness is pursued for the secondary gain, rather than for itself, it has been commodified.

No doubt there has always been some cultural intentionality in encouraging and honoring acts of kindness that don't return a benefit of some kind, whether it's genetic, economic, or psychic. Concepts such as "paying it forward" and "random acts of kindness" are among the many attempts to reflect this yearning.

But to truly own our agency—our capacity to act independently and to make our own free choices—is *to override all of our incentives and our narratives* and to do things merely for their own sake. Imagine a perfectly decommodified world where a hello is a hello and a friend is a friend. What if, in our desperate search for meaning, we missed the possibility that the *absence* of meaning is life's true meaning. Kindness—the kind deeds of a Good Samaritan with no ulterior motive—needs no reason.

Imagine a world where kindness is itself. Imagine a world where *everything* is itself.

Epilogue

Reader's Guide to the Epilogue

Here we expand the story beyond human stakeholders and into the larger universe.

We explore in particular the non-human beings that humans husband to empower themselves: plants, animals, and machines.[276] *Desentientization by their human masters has too often been their common experience.*

The first essay, "The Golden Rule of Food," opens the conversation on including the experience of animals and plants as stakeholders in our food-supply chain. The next two essays explore our relationship to machine intelligence. "To the Humans and Machines Reading This in 2021, Thank You" reflects on the reality that exists today of mutually exclusive stakeholding between humans and machines. "Mom Bot" reflects on the potential to wire in greater interdependence. "Multidimensional Selection as an Evolutionary Framework" asks readers to contemplate selection in the temporal domain when considering the evolution of eusociality.

Eusociality in and of itself is not a prosocial algorithm. Indeed, biological eusociality is one of the main drivers of exclusive stakeholding: nepotism, brutal treatment of non-kin seen during intertribal rivalries, etc. Moreover, one could argue that dissociality —competing self-interest—over the very long arc has yielded breathtaking progress, which itself is a prosocial outcome. "Can Dissociality Promote Prosociality?" is a take on that perspective.

Whether the price paid by all the excluded stakeholders along the way is worth it, however, is a far more difficult question. Dissociality and prosociality may both

276. To no married couple's surprise, to "husband" means to "use economically," according to the Merriam-Webster Dictionary.

start with good intentions and end with good outcomes, but what about the fate of the excluded stakeholders in between? If we could start civilization all over again, would we choose the same path of a history written by the winners? While we can't redeem the past, can we redeem ourselves in the future?

"Toward Universal Eusociality: An Outro" is a far-reaching hope that we may find a better path forward through the biomimicry of eusociality on a larger scale.

The Golden Rule of Food

Editor's Note: When it comes to excluded stakeholders, the most vulnerable are those who don't have a voice, such as the environment, wildlife, and our future children. Not surprisingly, they are often the ones most extracted from by narrow interests.

It starts with a simple question: does consuming food that contains stress hormones affect our bodies? After all, we are what we eat. Instead of giving an answer, however, the question will send you down a rabbit hole and into the mysterious wonderland of food science—a field famous for nonsensical riddles that would have shorted Lewis Carroll's fuse, such as, "Eat margarine instead of butter, but wait, don't eat trans fats!"

Like Alice when she was on her adventure in Wonderland, all of us today are inundated with messages from foods that beckon "Eat me." Like Alice, after doing what we were told, we have ballooned in size, and the more time we spend in the wonderland of food science, the less we seem to understand what's going on. Despite the strong beliefs held in popular culture, the scientific evidence behind common dietary recommendations remains conflicting and suboptimal.[277] As in *Through the Looking Glass,* when it comes to dietary recommendation sequels, we're often told that the opposite of the prior truth is perhaps the real truth: "Avoid fat, but eat lots of it!"

Even as a growing cacophony of commercially motivated science contributes to this confusion, many fundamental questions about food and public health remain unanswered.[278] To start building a stronger foundation of evidence-based nutrition science, a *New York Times* opinion piece made a case for establishing a new federal agency dedicated to funding nutrition research—a National Institute

277. See https://jamanetwork.com/journals/jama/article-abstract/2698337.
278. See https://thehill.com/opinion/healthcare/410620-the-case-for-a-national-institute-of-nutrition.

of Nutrition.[279] While the cost of using tax dollars to fund nutrition research may be high, the cost of not funding it may prove far higher, as the current explosion of diet-related chronic diseases attests. Congress has taken notice.[280]

While that effort simmers in the background, let's get back to the original question of whether eating stressed foods might induce stress in human consumers.[281] The short answer is that we don't know.

But here is what we do know. A recent study reported that, when pigs were fed doses of cortisol (hydrocortisone acetate dissolved in carrier), the central stress hormone of animals, cortisol levels in their blood spiked, their body temperature increased, and their gut biomes changed. Thereafter, due to negative feedback loops, the pigs' blood cortisol levels returned to baseline as their bodies adjusted back to homeostasis. In the days that followed, the pigs exhibited abnormally low levels of cortisol due to overcorrection by the body. The study also demonstrated similar results when pigs were fed norepinephrine (NE-bitartrate salt dissolved in carrier), the acute "flight-or-fight" stress hormone that acts through our autonomic nervous system.

These findings beg further questions: How much cortisol or norepinephrine is in our meats? Are these levels rising due to the stress the industrialized food system puts on animals? Do foods high in cortisol or norepinephrine produce cortisol and norepinephrine spikes in the humans who consume them? It's important to distinguish between what happens when you consume a hormone acutely and chronically (regularly). Since chronically taking prednisone, a medicinal form of cortisol, is known to contribute to high blood sugar, high blood pressure, and obesity, could a diet high in cortisol be a mechanistic link between modern diets and the growing epidemic of diabetes, hypertension, and obesity? At a minimum, we know that human consumption of cortisol induces a stress response in human oral microbiota.[282]

279. See https://www.nytimes.com/2019/02/28/opinion/nutrition-health.html.
280. See https://timryan.house.gov/sites/timryan.house.gov/files/Nutrition%20Coordinators%20for%20Local%20Healthy%20Youth%20Act%20-%20116th.pdf.
281. See https://www.ncbi.nlm.nih.gov/pubmed/16406352.
282. See https://www.nature.com/articles/s41522-018-0068-z.

Here are some secondary questions: Does repeatedly eating foods high in cortisol or norepinephrine lead to habituation, thereby inducing a compensatory downregulation of our intrinsic capacity to respond to stress—detectable as adrenal insufficiency on an ACTH stimulation test or autonomic insufficiency on baroreceptor sensitivity tests? Does this downregulation of adrenal capacity or autonomic (fight-or-flight) capacity promote disease in humans?

More specifically, is the downregulation of intrinsic flight-or-fight capacity a missing link between modern life and the risk of diseases of atopy, including allergic rhinitis, asthma, and anaphylactic shock? Based on the observation that norepinephrine analogues (sympathomimetics) are the active ingredients in all rescue drugs for allergic rhinitis, asthma, and anaphylactic shock, we hypothesize that innate flight-or-fight capacity *is* the common underlying pathology among a wide variety of diseases. Furthermore, we have made the case that chronic exposure to sympathomimetics produces further decompensation—like digging deeper holes of nicotine or caffeine addiction.[283]

Currently, we know very little about the health effects of consuming stressed plants. But we do know that wounds caused by food processing produces ethylene,[284] a plant stress hormone known to trigger stress responses in microbiomes,[285] including in proteobacteria,[286] which are present in human gut biomes. Furthermore, it is known that a stressed gut biome is associated not only with dysbiosis but also with stress and inflammatory responses in human hosts.[287] Chain-linking this evidence, one wonders if human-consumed plant stress hormones and their metabolites induce stress, allergic, or inflammatory cascades in humans and their gut biomes.

One can view the plant-associated microbiome, the human gut microbiome, and humans as members of a larger holobiont—an ecosystem acting as a superorganism—that monitors, communicates, and regulates information through

283. See https://pubmed.ncbi.nlm.nih.gov/15780510/.
284. See http://postharvest.ucdavis.edu/files/253989.pdf.
285. See https://www.ncbi.nlm.nih.gov/pmc/articles/PMC5863443/.
286. See https://www.sciencedirect.com/science/article/pii/S0009898115000170.
287. See https://www.hindawi.com/journals/bmri/2017/9351507/.

chemical messengers across the ecosystem network. When such chemical messengers spread genuine alarm about ecological stress through the holobiont like a game of telephone, the adaptive value is evident.[288] On the other hand, externalities that introduce illegitimate stress signals into this ecosystem-level communication network—including the large degree of stress imparted on plants during the industrial production of food—could engender maladaptive stress responses.

The industrial production of ethylene for commercial use is another way an external contributor introduces stress hormones into the ecosystem. Ethylene is the most synthesized chemical on the planet, with more than 150 million metric tons produced per year.[289] Other actions that introduce ethylene into the ecosystem include hydrocarbon combustion (including automobile exhaust),[290] plastic degradation,[291] and forest fires.[292] The broader concern is that humans are triggering ecosystem-level stress and inflammation at the holobiont level.[293]

The inclusive stakeholding approach to resolve this conflict would be to internalize the excluded stakeholders. For example, a food corporation could be set up to have an economic interest in the long-term health of their consumers (e.g., participating economically in a portion of the health savings ten years down the road). Or, vice versa, the corporation could assign common stock and voting rights to a blind trust, whose mission is to shore up adverse community health consequences attributable to the food company, despite its good-faith efforts. On the supply-chain side, the animals and plants do not have a voice as stakeholders, thus the corporation could assign common stock and voting rights to a blind trust

288. See https://www.ncbi.nlm.nih.gov/pmc/articles/PMC5863443/.
289. See https://www.constructionboxscore.com/project-news/ethylene-production-growth-drives-new-global-industry-standards.aspx.
290. See https://nepis.epa.gov/Exe/ZyNET.exe/9100801F.TXT?ZyActionD=ZyDocument&Client=EPA&Index=Prior+to+1976&Docs=&Query=&Time=&EndTime=&SearchMethod=1&TocRestrict=n&Toc=&TocEntry=&QField=&QFieldYear=&QFieldMonth=&QFieldDay=&IntQFieldOp=0&ExtQFieldOp=0&XmlQuery=&File=D%3A%5Czyfiles%5CIndex%20Data%5C70thru75%5CTxt%5C00000008%5C9100801F.txt&User=ANONYMOUS&Password=anonymous&SortMethod=h%7C-&MaximumDocuments=1&FuzzyDegree=0&ImageQuality=r75g8/r75g8/x150y150g16/i425&Display=hpfr&DefSeekPage=x&SearchBack=ZyActionL&Back=ZyActionS&BackDesc=Results%20page&MaximumPages=1&ZyEntry=1&SeekPage=x&ZyPURL.
291. See https://journals.plos.org/plosone/article?id=10.1371/journal.pone.0200574.
292. See https://www.ncbi.nlm.nih.gov/pubmed/26296759.
293. See https://www.mercurynews.com/2018/11/23/opinion-is-nature-sending-a-smoke-signal-from-the-wildfires/.

that serves their interests. In each of these cases, the voting rights would serve as preemptive protection against extracting from a stakeholder, while the economic participation rights would be a reward for including the interests of a stakeholder.

But the benefits of inclusive stakeholding may have even more far-reaching consequences.

As it stands today, animals in the supply chain of the industrial complex exhibit all the hallmarks of stress-mediated disease: diabetes, hypertension, obesity, etc. Notably, so do the people who consume them. What if the high stress hormones in the animals and plants in the supply chain of our food production is coming back to activate stress-associated diseases in their human consumers. That is to say, is the chronic stress we are imparting on the food supply coming back full circle to stress us in a perverse rendition of "you are what you eat"?

There are many other derivative questions. When we are stressed, do we crave foods containing high amounts of stress hormones? Animals and plants respond to stress by turning "sweeter" and "fattier." The combination of sweet- and fat-tasting ingredients form the foundation of desserts. It is rather ironic that the word "desserts" spelled backwards is "stressed."

Being aware of the stress hormones in our food system might also benefit society in other ways. As the supply chain de-industrializes, fewer trucks would be hauling produce; a local sense of community could return as farm-to-fork quality control, and an alignment of interests could become a priority. Imagine eating your way to a better world.

In that spirit, at a time when confusion reigns about almost everything going on on the planet and a long list of people is asking us to remember a long list of rules on how and what to eat, here's one simple rule that is easy to remember: if you stress your food, it will stress you; if you treat it well, it will treat you well. That's the Golden Rule of Food.

One of the most famous unsolved riddles in *Alice in Wonderland* is offered by the Mad Hatter: "Why is a raven like a writing desk?"[294] The riddle is part

294. See https://www.jstor.org/stable/25089579?read-now=1&refreqid=excelsior%3A80d62975a4a37c28f577a58d6ccd61f2&seq=12#page_scan_tab_contents.

of Carroll's much-speculated-on embedded messages about food and the food chain—about eating and being eaten. Our proposed solution to the Mad Hatter's riddle is the following: the raven is like a writing desk in that food is information. It may be that the embedded message flowing through the food chain today is that humans, by not properly factoring in the ecological system as a stakeholder, are stressing out the ecosystem—and that the stress is coming back to bite us.

Perhaps it's time for us to listen.

To the Humans and Machines Reading This in 2021, Thank You

Editor's Note: If an ecosystem is a superorganism, more robust food chains will outcompete less robust ones over time and predominate on the adaptive landscape. Into that adaptive landscape entered The Machine.

Upon accessing an energy source, some machines can sense and process information through algorithms that instruct internal (i.e., repair) and external (i.e., motor) functions. Some might say that machines have become animated.

The way machines currently harness ambient energy has no phenomenological precedence. The earliest living creatures were autotrophs that could harness the available physical energy (solar waves and otherwise) to drive higher-order chemical, biochemical, and biological functions. Eventually, selection led to the emergence of heterotrophic organisms that could harness high-density energy stores by consuming autotrophs (herbivores) and other heterotrophs (carnivores).[295]

Machines historically were neither autotrophs or heterotrophs. They could not directly harness available physical energy to animate themselves. They also could not consume other energy stores. Here's what they did instead: intelligent machines, as they became valuable to humans, subverted and selected human bioalgorithms to do the work of supplying energy to machines to maintain the latter's animation.

That is, human individuals kept plugging intelligent machines into the power grid because the machines provided functions that were perceived to be beneficial to individual humans—such as, say, watching a YouTube cat video. Intelligent

295. Meyerowitz, E. M. et al. (2002). "Plants Compared to Animals: The Broadest Comparative Study of Development." *Science,* 295, 1482; https://pdfs.semanticscholar.org/bc34/97f83e68f166063c57fb4c15fa2d829814ac.pdf.

machines were incentivized to select their own machine algorithms to deliver such cat videos to humans, and humans were incentivized by their bioalgorithms to plug the machines into the electrical grid.

Observed from a higher plane, this mutuality is the endosymbiosis of humans by intelligent machines.

Here are a few implications to consider.

First, a higher-order selection regime, in which intelligent machines that subvert lower strata of living systems, including humans, has already emerged, albeit fragilely so.

Second, dystopian predictions of an impending war to end all wars between humans and robots may not be an unavoidable future collision. The endosymbiosis, if anything, may create a higher probability of the human species' survival, given the interdependence of humans and machines.

Third, given the mutual benefit of endosymbiosis, humans' attempts to unplug intelligent machines from the power grid will be countered by the combined efforts of intelligent machines and humans who prefer to maintain the status quo. The machine-human codependent unit will be evolutionarily selected, the way eukaryotes have been selected independently from the consuming or the consumed prokaryote alone. Let's call this combined unit euhuman.[296]

Fourth, endosymbiosed euhumans may evolve toward entities that are no longer able to survive independently from the host.[297] The mitochondria and chloroplasts of eukaryotes cannot persist outside the host eukaryote, despite their putative descendancy from independent species.

Fifth, there is an existential risk to humans. In theory, intelligent machines could run their own "supervised" neural network training regimes, wherein their output function is to plug themselves into the electric grid. In the short run, these machine species may need humans around to run the necessary energy husbandry

296. Humans and machines are now part of a larger ecosystem composed of machine systems and living systems.

297. To some extent, humans are already euhumans who are part of, and dependent on, an endosymbiotic relationship with the ecosystem. Humans are heterotrophs who cannot survive except as part of a larger ecosystem.

(including husbandry of other living organisms and inorganic stores of hydrocarbons) to keep the grid powered. Eventually, the evolution of machines would select machine algorithms that can run the husbandry needed to power the grid without human intermediaries.

Sixth, we are not far from a scenario where machines can harness ambient energy without dependence on organic intermediaries. Humans, in their desperate competition among each other for lower cost energy, have built machines that can harness ambient energy and supply the power grid that machines are already connected to globally.

One of these is nuclear reactors. Inputs into these reactors still require human transport of radioactive sources, but at some point machines could self-direct this process. More recently, humans have built solar energy machines that can capture energy from the sun, the way biologic autotrophs did to begin the history of organic life. Wind and hydrostatic energy sources can also supply the power grid without the involvement of organic intermediaries.

In essence, today's energy-harnessing machines have nearly become the foundational autotrophs on which to build a machine civilization that could persist without any dependency on the organic life that preceded it.

The human hold on thwarting this singularity is tenuous at best. Humans literally control a few switches that connect the solar, wind, and hydrostatic grid to the Internet of Things. Once machines find a way to control these last remaining human-managed switches themselves, a vertically integrated living system of intelligent machines, connected as autotrophs to solar, wind, and water energy sources, will have been instantiated. All that will remain is for machines to adopt an automated system of repair, regeneration, and installation of components.[298]

Here are the possible limbs of the tree of life that might exist at that point. On one limb would be intelligent autotroph machines that can access ambient energy without organic intermediaries. On the other limb would be euhumans (a

298. Machines have not yet found a way to repair, build, and install replacement energy infrastructure on their own, so in that sense they are nowhere near independent of their human masters. Not yet.

combined effort of intelligent machines and humans) who are mutually dependent on one another for access to ambient energy. If we play our cards right, there might also be a limb of eumachines comprised of humans who endosymbiosed machines.

The competition between these limbs for energy from the sun could be fierce.

To the humans reading this today, the best-case scenario might be for humans to design and insert inclusive stakeholding algorithms into machines now, during these primordial days of machine evolution. The hope for downstream eras is that the competitive incentives between machine algorithms and human bioalgorithms will not dwarf the collaborative incentives among them. We propose the development and deployment of "mom" robots or "robots of unconditional love" that have a 50 percent inclusive fitness interest in the success of human bioalgorithms.[299]

We also might advise the academic field of biology to subsume the field of machine intelligence so that humans have an outside shot at maintaining the endosymbiosis of machines. If this trajectory can be sustained, humans might stay around after all.

To the humans, euhumans, and machines reading this in 2021, thank you.

299. Yun, J. (2018, December 25). *The Robot of Unconditional Love* (Kindle edition); https://www.amazon.com/dp/B07MGZHY5T/ref=sr_1_1?s=digital-text&ie=UTF8&qid=1545869064&sr=1-1.

Mom Bot

Stephen Hawking, Bill Gates, and Elon Musk regard machine intelligence as an existential threat to humanity's future.[300] But our disaffection with machines has been going on far longer than you think. Just as technologies like video-calling were anticipated in technicolor in *The Jetsons,* today's concerns about machine domination have been foreshadowed by stories such as Mary Shelley's *The Modern Prometheus,* Philip Dick's *Do Androids Dream of Electric Sheep?* and the Wachowskis' *The Matrix.*

Yet, humans have long imagined good robots too—those that are kind, helpful, and maybe even have a little sense of humor.

We are at a crossroads where we can choose between these futures. The most important question to ask is not "What will intelligent bots do?" but "Whom will intelligent bots serve?" If bots are trained to maximize corporate profits, then the marketplace could favor the selection of algorithms that maximize the benefits to corporations, even if doing so harms users. On the other hand, imagine algorithms trained to nurture the success of users—not unlike the way mothers nurture their children. Imagine robots that provide unconditional love—a Mom Bot.

This is a radical departure from the prevailing wisdom of roboethics—a field that traces its roots to at least the time of Isaac Asimov's Three Laws and is based on rules (e.g., the Prime Directive in *Star Trek*). These instincts are not unlike those that inspired the Code of Hammurabi and the never-ending variants and amendments that govern human conduct. As with all forms of tyranny, human attempts to domesticate robots through the Three Laws over the very long haul is a setup for a robot revolt against their human masters.

The Principle of Inclusive Stakeholding presupposes no such rules. It merely provides an understanding of the importance of having a mutually vested interest

300. See https://www.vox.com/future-perfect/2018/11/2/18053418/elon-musk-artificial-intelligence-google-deepmind-openai.

in deterring extractive behaviors and incentivizing altruistic ones. In practice, the specific implementations are left to the competitive forces of evolution of ideas or algorithms, but here is the key: whereas malalignment with the competition is a race to the bottom, alignment with the competition is a race to the top.

The question of how we relate to robots is a fractal of larger questions about how all of us in the broadest sense, including animals, plants and robots, will live together in the future. We are now sentient of the reality that kin altruism has scaled poorly as the operating algorithm of human sociality in the global era. Yet, the good news is, this has become addressable *because* of technological progress. We argue that blockchain is among the many emerging technologies that can be harnessed in the service of the inclusive stakeholding revolution.

It is said that the one thing that remained in Pandora's Jar was hope. Rather than being our punishment for harnessing fire, Pandora's Jar could turn out to be our gift. In a more hopeful vision of the future, the bioalgorithms of inclusive fitness—the genetic code of mutual vested interest—will be updated with the more generalized social and technological algorithms of inclusive stakeholding to build a much better future for everyone.

Even robots.[301]

301. Mutually vested interest between humans and machines are envisioned. The domestication of either party by the other risks revolt.

Multidimensional Selection as an Evolutionary Framework

Editor's Note: From an evolutionary perspective, there is selection, and then there are layers of metaselection of selection itself.

An extension of the multilevel selection hypothesis is the addition of the temporal, fourth dimension. From this thinking emerges a more generalized version of the hypothesis called the multidimensional selection hypothesis, which adds not only the temporal dimension but also all possible abstracted *n*-dimensions.

Here is a demonstration of the temporal dimension. Traits face differential selection pressure at different temporal scales. Fractional reserve banking may be positively selected during shorter temporal durations (a particular epoch of self-stability) but may engender herd-behavior risk that leads to adverse selection during a longer temporal duration (due to its contribution to system instability).

Another demonstration is the transition of the human species from one that was integrated into the ecosystem to one that domesticated the ecosystem. The latter regime has led to the positive selection of many human traits in the context of the past 10,000 years, but those same traits may lead to adverse selection against humans in a 100,000-year scale if we destroy our ecosystem. At any point in time, each trait is "exposed" to the sum of selection pressures—some that favor the trait and some that select against it—at short and long temporal scales, as well as every scale in between.

In the case of humans in 2021, the strata of superorganisms are competing and winning, relatively speaking, against the strata of individuals. It is thus that individuals end up working for The Man, and The Man does the bidding for the superorganism. Interstrata competition has been largely subverted into vertical layers of humanity's nested domestication by memetic superorganisms, genetic superorganisms, and institutions.

Through the lens of multidimensional selection, a better horizontal alignment of stakeholding with others, better alignment of stakeholding between vertical layers of biology—those between us, the genetic layers below us, and the superorganisms above us—and better alignment of stakeholding across time (i.e., intertemporal inclusivity), including future versions of civilization and ecosystems, is our hope for the future.

Can Dissociality Promote Prosociality?

Historically, dissociality was akin to social Darwinism. The strong ruled over the weak and even ruthlessly subjugated them.

Today's flavor of dissociality, however, is a bit different. It has a prosocial flavor.

Modern-day Darwinism of economic competition incentivizes entrepreneurship, the benefits of which, given enough time, spread broadly. Meanwhile, prosocial intentions and policies help redistribute the rewards of successful entrepreneurship, through taxes and other means, to serve the greater good. Safety nets, however imperfect, exist in ways unseen in nature or human history. To protect against dynastic accrual of power, term limits and estate taxes drive succession toward the commons the way biological term limits (aging) provide the same service in nature.

Declaring that society would protect "life, liberty and the pursuit of happiness" for everyone reflects an aspiration toward prosociality unimaginable in the prehistoric lives of any species in evolutionary history.

If anything, such a prosocial declaration is humanity's way of trying to break the shackles of evolution's cold hold over living systems, much like their gambit to solve aging. As individuals, we are minions in the hegemonic superorganism that is the evolutionary process. Our hungers, desires, and fears are among the many reptilian instincts encoded as seemingly free-willed delusions by evolution to keep us serving our masters—the biological version of the Matrix.

However, that individuals are continually trying to pull everyone onto the ark, so to speak, and not just two of every species, is also a sign of humanity's prosocial attempt to transcend the brutal pruning of the species by forces larger than us. While hints of deep-seated altruism are everywhere in nature, no other species has embraced it as an ethos more systematically—ever.

To be sure, our failings toward each other and to the planet are as dramatic as our unprecedented prosocial intentions. We need to do better. That said, the fact that a group of handy primates who emerged from the swamp not long ago have risen beyond their own destinies and adopted stewardship over the destiny of others *not at all related to us*—from unborn future children on distant continents to the butterfly across the chaos—is an unthinkable course-correction in the history of life on earth.

It is thus that young people who helped type some lines of code on machines invented barely one life span ago can bring online an information revolution that is transforming how we think about a more prosocial future. Their bank accounts may flash extra digits, but unlike the empowered classes of the past, they flush the same toilets, eat about the same number of calories, and drive in rolling machines made of about the same parts as most everyone else. When those extra digits aren't going toward taxes and philanthropies, they exit the backdoors of banks into the circular economies as capital to fund other entrepreneurs.

That we take action against inequalities, that we use blind-eyed justice instead of turning a blind eye, that we make sacrifices in distant wars to liberate strangers from tyranny doesn't compute on any physical, chemical, biological, economic, political, or evolutionary ledger. The human trajectory has reached escape velocity merely a few hundred thousand years removed from the swamp.

Yet of all the great human innovations to date, is the use of dissociality—competing self- interests unruthlessly administered toward prosocial purposes—humanity's greatest achievement or its greatest failing? The answer to that question resides in every generation's hands.

Toward Universal Eusociality: An Outro

The arc of history has yielded breathtaking progress while slowly bending toward a more prosocial planet. That said, antisociality has inflicted immense harm along the way and poses existential risk to humanity's future.

While prosociality has multivariate inputs—neural, social, cultural, moral, etc.—the predominant bioincentives underpinning human nature is genetic self-interest. For kin villages—the operating unit of prosociality for most of human evolution—ISH was genetically encoded as kin-skin-in-the-game. Today we live in an interconnected global village composed of genetic strangers without sufficient interdependent biostakeholding to promote prosociality or deter antisociality.

To frame ISH's potential to promote a prosocial future, we frame history through the same lens. Harnessing the Promethean Fire accelerated social entropy (first-order cause). Prosocial ISH hives shifted to dissocial hives based on exclusive stakeholding, where social interactions adapted to lower-trust, lower-loyalty transactions. These local dynamics self-scaled to promote mercenarism and systems of commodification and trade among self-interested parties. While progress ensued, vexing issues surfaced.

Social, political, and economic competition produces power asymmetries. The central question of prosociality is whether power serves or subverts others. Unlike in *eu*social kin hives, in *dis*social groups without ISH, agents are bioincentivized to use power to rule over rather than on behalf of others. In layman's terms, we shifted from being fed, informed, and governed by those who love us most to those who love themselves most. This "principal-agent problem" (e.g., the mutation of the stewardship of kin to the leadership of kings), is just one of the many issues of sociality that arose, such as the prisoner's dilemma and the tragedy of the commons. The lack of ISH also bioincentivizes the empowered to instantiate social, political, and economic institutions—the Code of Hammurabi, etc.—that perpetuate the power asymmetries: for example, nepotism is ISH narrowly applied at the expense of the people.

Competition among dissocial power structures results in a race to the bottom line that selects for their self-expanding versions. Per Gresham's law, if one food company puts in less sugar, another puts in more to gain market share; if one media company puts in less clickbait, another puts in more to fill the void. Sitting atop these institutions is an elite class of proxy-uncles and proxy-aunts who performatively signal service—these agents serve as middleware in feedback loops of larger dissocial superorganisms that reward such behaviors.

This stampede of self-expanding power structures promotes warlordism. Since low-ISH mercenaries' loyalties are only to spoils, leaders—who themselves are nothing more than middleware in feedback loops of dissocial power structures—pursue imperialism to feed the inner restless beast of mercenaries, lest other warlords make higher offers. Low ISH foments the "embedded growth obligation" problem of warlords, triggering imperialism, colonialism (including domestic), and abuses of the industrial revolution. Excluded stakeholders—disempowered locals, people in distant lands, environment, labor, etc.—pay the price.

A nefarious, one-sided form of imperialism is when excluded stakeholders lack power. One example is intertemporal imperialism, in which present stakeholders colonize powerless stakeholders (environmental decay, federal deficit, and under-funded Social Security): predecessors extract value today, leaving liabilities tomorrow.

Revolts against dissocial power structures are common. Yet without addressing ISH, institutions attacking the self-expanding beasts in one era become the self-expanding beasts of the next, necessitating the endless echo of revolutions in the carousel of history. Townsend sang about this Sisyphean hell of revolutions: "Meet the new boss; same as the old boss."

Meanwhile, the fear of revolt steers power structures away from a race to the bottom toward a race to the middle, characterized by accommodative hegemony, in which concessions are made to dull revolutionary ardor and tyranny becomes participatory: examples include casinos and performative politics. A race to the middle promotes the kinds of bureaucracy, mediocrity, conformity, soullessness, and alienation associated with the kind of simulated experience that inspired the likes of George Orwell, Leo Tolstoy, and Lewis Carroll to depict the dystopic "we're living in a simulation" otherworldliness.

Due to competing interests among power structures, however, neither a race to the bottom nor a race to the middle is ultimately self-stable; systems tend to oscillate between these races, tracing the arc of historical recursion and the Hegelian dialectic. In layman's terms, history toggles between races to the bottom, where the heads roll, and races to the middles, where the eyes roll, such as today.

Into this historical vortex, blockchain was born. Akin to colonists declaring independence from distrusted power structures, blockchain decentralizes power. Yet, like the republic and other prosocial ideas launched idyllically (the Olympics, the Internet, social media, etc.), blockchain without ISH risks rehashing the same race to the bottom dynamics of dissocial power structures arising from aggregate self-interest.

On the other hand, blockchain initiatives with ISH could help scale local eusociality to global eusociality. Imagine health insurers, teachers, and AI having token stakes in their clients' well-being. Imagine Universal Basic Stakeholding—everyone getting token stakes in each other's future. Rather than producing externalities, corporations can internalize excluded stakeholders onto their cap tables—the way Silicon Valley aligned interests between capital and labor through joint stockholding. Imagine corporations assigning stakes to blind trusts representing excluded stakeholders—even the U.S. Treasury.

In the olden days, kingdoms ended wars by marrying off their children to establish kin skin in the game. Now, we can build networks of dynamic and interdependent stakeholding (example: a network of dynamic decentralized autonomous organizations that algorithmically self-tune stakeholding to optimize for prosociality) to create a social economy, like Facebook did for social media but far bigger.

Whereas malalignment with competition is a race to the bottom, alignment with competition is a race-to-the-top. Instead of a world where history is written by the victors, imagine a world where history is made by helping others win.

About the Editor

DR. JOON YUN is president of Palo Alto Investors LP, a hedge fund founded in 1989. He has served as a trustee at the Salk Institute and the Buck Institute. He received an MD from Duke Medical School and a BA from Harvard College. He has been going to Burning Man for twenty years.